AF840757

HAZARDOUS MATERIALS IN THE SOIL AND ATMOSPHERE: TREATMENT, REMOVAL AND ANALYSIS

HAZARDOUS MATERIALS IN THE SOIL AND ATMOSPHERE: TREATMENT, REMOVAL AND ANALYSIS

ROBERT C. HUDSON
EDITOR

Nova Science Publishers, Inc.
New York

Copyright © 2006 by Nova Science Publishers, Inc.

All rights reserved. No part of this book may be reproduced, stored in a retrieval system or transmitted in any form or by any means: electronic, electrostatic, magnetic, tape, mechanical photocopying, recording or otherwise without the written permission of the Publisher.

For permission to use material from this book please contact us:
Telephone 631-231-7269; Fax 631-231-8175
Web Site: http://www.novapublishers.com

NOTICE TO THE READER

The Publisher has taken reasonable care in the preparation of this book, but makes no expressed or implied warranty of any kind and assumes no responsibility for any errors or omissions. No liability is assumed for incidental or consequential damages in connection with or arising out of information contained in this book. The Publisher shall not be liable for any special, consequential, or exemplary damages resulting, in whole or in part, from the readers' use of, or reliance upon, this material.

Independent verification should be sought for any data, advice or recommendations contained in this book. In addition, no responsibility is assumed by the publisher for any injury and/or damage to persons or property arising from any methods, products, instructions, ideas or otherwise contained in this publication.

This publication is designed to provide accurate and authoritative information with regard to the subject matter cover herein. It is sold with the clear understanding that the Publisher is not engaged in rendering legal or any other professional services. If legal, medical or any other expert assistance is required, the services of a competent person should be sought. FROM A DECLARATION OF PARTICIPANTS JOINTLY ADOPTED BY A COMMITTEE OF THE AMERICAN BAR ASSOCIATION AND A COMMITTEE OF PUBLISHERS.

Library of Congress Cataloging-in-Publication Data

Hazardous materials in the soil and atmosphere : treatment, removal, and analysis / Robert C. Hudson (editor).
p. cm.
Includes index.
ISBN 1-60021-250-6
1. Soil Pollution. 2. Air--Pollution. 3. Soil remediation. 4. Air quality management. 5. Hazardous substances--Environmental aspects. I. Hudson, Robert C. (Robert Charles), 1972-
TD878.H39 2006
628.5--dc22 2006016749

Published by Nova Science Publishers, Inc. ✦ New York

CONTENTS

Preface vii

Chapter 1 Combined Treatment of Styrene-Contaminated Soil by Rhamnolipid Washing Followed by Anaerobic Biodegradation **1**
Yuan Guo and Catherine N. Mulligan

Chapter 2 A Study on the Feasibility for Controlling Trace Metal Leaching Behavior from Municipal Solid Waste Incineration Ashes **39**
Wu-Jang Huang

Chapter 3 Alternative Halomethylation Processes and the Corresponding Preparation Routes for Anion Exchange Membranes by Avoiding the Use of Hazardous Materials Chloromethyl Methyl Ether (CME) and Bis-Chloromethylether (BCME) **55**
Tongwen Xu

Chapter 4 Chemical Fractions and Bioavailability of Heavy Metals in Polluted Soils **89**
Geeta Tewari, P.C. Srivastava, Bali Ram and Lalit Tewari

Chapter 5 Removal of Phenolics from Aqueous Solutions by Using Modified Clays As Adsorbents **117**
Saadet Yapar

Chapter 6 Thermolysis of Transition Metal Complexes: A Review **143**
Gurdip Singh, Durgesh Kumar Pandey and Chandra Prakash Singh

Index **161**

PREFACE

Hazardous waste is a waste with properties that make it dangerous or potentially harmful to human health or the environment. Hazardous waste generally exhibits one or more of these characteristics: ignitability, corrosivity, reactivity or toxicity. The universe of hazardous wastes is large and diverse. Hazardous wastes can be liquids, solids, contained gases, or sludges. They can be the by-products of manufacturing processes or simply discarded commercial products, like cleaning fluids or pesticides. One major type is radioactive waste. This new book brings together the latest research in this diverse field.

Chapter 1 discusses Styrene classified in the EPA's Toxic Release Inventory (TRI) as a carcinogen. It enters the environment during the manufacture, use, and disposal of styrene-based products. The interaction of styrene with soil is a very complex phenomenon and means are required to understand this matter more fully. The objective of this research is to develop a system for the removal and biodegradation of styrene and to determine the mechanisms for styrene removal from soil by biosurfactant (rhamnolipid). The system will consist of two stages: soil flushing followed by an anaerobic biobarrier for groundwater treatment. In this research, a long-term styrene adsorption test was performed. The initial column test results showed that after 90 days, high concentrations (from 700 mg/kg to 4700 mg/kg) remained. A biologically produced surfactant named rhamnolipid (JBR-425) was used in the batch removal tests at a concentration of 0.5%. From the preliminary results, 72.7% of the styrene was removed from the soil after a day and 88.3% after 5 days (original styrene concentration was 32,750 mg/kg). From a 16,340 mg/kg styrene contaminated soil, 90% of the styrene was removed after a day. In the styrene biodegradation tests, the initial results showed that with anaerobic biomass, styrene degraded 37.9% after day 1 and 67% after day 3. Both removal rates were higher than without the biomass (12.5% removal at day 1 and 46% removal at day 3). Also, the rhamnolipid did not inhibit styrene biodegradation. The initial results from the styrene removal and biodegradation tests showed that styrene could be removed from soil by the rhamnolipid and that styrene could be biodegraded under anaerobic conditions with the biomass.

In a modern municipal solid waste incineration (MSWI) plant, two air pollution control (APC) residue streams of semi-dry (SD) scrubber residue and fabric filter (FF) ash are generated and usually on-line combined together at a random mixing ratio before for collection, as discussed in chapter 2. And the collected combined ash is called as MSWI fly ash, in order to distinguish from the coal fly ash. An common phenomena has been recognized in a real MSWI plant is that the toxicity characterization leaching procedure

(TCLP) test of a collected MSWI fly ash has a high fluctuation in the leaching concentrations for several easy to leach elements, such as lead and zinc.

In the previous studies [1, 2], the authors have reported the chemical composition analysis of semi-dry (SD) scrubber residue and fabric filter (FF) ash from seven MSWI plants and coleaching behavior of lead from SD/FF mixtures from one of them. In addition, an empirical equation based on the major chemical compositions of original SD residues and original FF ashes has been proposed to describe the coleached concentration of lead from the mix ashes in any ratio [2]. In this study the application feasibility of equation has been extended to three series of synthetic APC residues with varying contents of FF ashes collected from three MSWI plants. And for the first time the fluctuation in TCLP concentration of lead has been suggested that it would be strongly depending on the mixing ratio of FF ash to SD stream mixture, basing on the prediction of our proposed empirical equation. Our data also indicates that the lead (II) concentration of the synthetic APC residues in the TCLP test is lower than the regulatory level, 5.0-ppm, if the mixture contained FF ash less than 50-wt%. If MSW incinerator plant operators blend APC residues with less than 50-wt% of FF ash to consistently produce a non-hazardous and recyclable fly ash under TCLP test, less than 50-wt%FF ash in a collected MSWI fly ash.

Chapter 3 reviews ion exchange membranes, that have been widely used in various industrial fields, such as electrodialysis to concentrate or deionize aqueous electrolyte solutions, diffusion dialysis to recover acid or alkali from waste acid or alkali solutions, anion exchange layer constructed into a bipolar membrane for cleaning productions and separations, polymer electrolyte for fuel cell, etc. To satisfy with these requirements, strong acid type cation exchange membranes with sulfuric group and strong base type anion exchange membrane with quaternary ammonium group were commonly prepared from styrene-divinylbenzene copolymers. In such a route, preparation of a cation exchange membrane seems to be easier by simple sulphonation with concentrated sulfuric acid or chloride sulphonic acid; while preparation of an anion exchange membrane had to experience more complicated procedures: chloromethylation and quaternary amination. Particularly, in the chloromethylation, the used chloromethyl methyl ether or bis-chloromethylether is a potent carcinogen and its use has been restricted since the 1970s. Till now, some efforts have been made to cancel the use of such hazardous material. Therefore, in this chapter, these routes for anion exchange membrane preparations in the absence of chloromethyl methyl ether or bis-chloromethylether will be reviewed and commented.

In a greenhouse experiment, the effect of different levels of heavy metals applied through inorganic salt sources or enriched organic sludge on the dry matter yield and total uptake of heavy metals by spinach (*Spinacia olerecea* L cv "All green") was studied in three different soils in chapter 4.

Treated soils were also analyzed for pH, 1M $CaCl_2$ – and 0.005 M DTPA- extractable heavy metals and five different operationally defined chemical fractions of heavy metals: F_1 (Soluble + exchangeable, in 0.1M $Sr(NO_3)_2$), F_2 (carbonate bound in 1 M NaOAc of pH 5.0) , F_3 (Oxidizable fraction in 5% NaOCl of pH 8.5), F_4 (Reducible fraction in 0.2M oxalic acid + 0.2M NH_4 oxalate + 0.1M ascorbic acid , pH 3), F_5 (Residual fraction in (HF – $HClO_4$ digest). Except Cd, greater proportion of all heavy metals was found in Fe-Mn oxide bound and residual fractions in all polluted soils. In case of Cd, NaOAc extractable fraction was the highest among all fractions while water soluble and exchangeable fraction was higher in inorganically polluted acidic soil. As compared to inorganic salts, sludge application increased the more labile water soluble and exchangeable Zn in all soils. Inorganic salt application increased water soluble and exchangeable Cd, Cu, Ni and Pb; especially in acidic soil. Water soluble and exchangeable Cd, Cu, Ni and Zn and residual Pb fractions showed significant negative correlation with the soil pH.

Heavy metals applied through inorganic salts drastically reduced the dry yield of spinach, especially in acidic soil. The highest uptake of all heavy metals by spinach was recorded in neutral soil. Total uptake of heavy metals applied through sludge treatment was manifold higher than inorganic salt treatment. The 1M $CaCl_2$ extractable amount of Cd, Cu and Ni showed significant positive correlation with the concentration of these heavy metals in spinach tissue. The 0.005M DTPA extractable Cd, Cu, Ni and Zn showed significant positive correlation with the concentration of these heavy metals in spinach tissue. The content of 0.005M DTPA extractable Ni and Zn also showed significant positive correlation with total uptake of these metals by spinach plant.

Phenolic compounds are priority pollutants due their toxic and harmful effects even at low concentrations. In addition to the biological sources such as olive oil mill effluents and animal wastes; the sources of phenol contamination are mainly the petrochemical, coal conversion and phenol producing industries, as discussed in chapter 5. There are many methods for the removal of phenol from aqueous solutions, adsorption seems to offer a better prospect than the others, which require high pressure and /or temperature, and complex equipment. Since synthetic adsorbents are relatively expensive, the use of natural adsorbents is an active area of research. Additionally, some biological materials and industrial wastes are also used in the removal of phenol. Among the naturally occurring adsorbents, clay has the most common application due to its outstanding adsorption properties. However, the modified forms of clay are generally used in the removal of phenolics since the negatively charged and hydrophilic surfaces of the mineral render it to be less efficient. The modified forms are obtained through calcinations, pillaring, acid activation, and the replacement of exchange cations with quaternary alkylammonium derivatives. The adsorption properties of the modified forms are generally tested by batch experiments. Application of the results to industrial scale requires the determination of kinetic and isotherm models. Langmuir and Freundlich models are widely used in the determination of the adsorption behavior of phenolics on modified surfaces. In terms of design considerations, the model proposed should be as simple as possible and contain minimal parameters. This article will cover a literature review on the removal of phenolics by using modified clays. The methods used in the modification and the adsorption properties of the modified forms will be summarized. The forms of isotherms and isotherm equations used in the modeling of adsorption behavior will also be given.

In chapter 6 thermal behaviour of various transition metal complexes have been reviewed thoroughly. Thermolyses of complexes are studied by TG, DTG, DTA and DSC etc. and pathways of decomposition have been proposed. Any phase transition during thermolysis, studied by the techniques known as 'high temperature reflectance spectroscopy' (HTRS) and 'dynamic reflectance spectroscopy' (DRS) have also been discussed. Intermediates formed during thermolysis has been investigated. Most of the complexes were found to decompose in multisteps. Effect of change of atmosphere on thermal decomposition has also been visualized. End products of thermolysis has also been discussed under oxidative, non-oxidative, quasi-isothermal and isobaric conditions. Stress has been given to compare the thermilytic pattern and thermal stability of the complexes with changing ligand or metal or counter anions. Attention has been given to investigate the kinetics and general mechanism of thermolysis of the complexes.

In: Hazardous Materials in the Soil and Atmosphere
Editor: Robert C. Hudson, pp. 1-38
ISBN 1-60021-250-6
© 2006 Nova Science Publishers, Inc.

Chapter 1

COMBINED TREATMENT OF STYRENE-CONTAMINATED SOIL BY RHAMNOLIPID WASHING FOLLOWED BY ANAEROBIC BIODEGRADATION

Yuan Guo and Catherine N. Mulligan[1]
Dept. Building, Civil and Environmental Eng. EV 006-187,
Concordia University, Montreal, Quebec, Canada

ABSTRACT

Styrene is classified in the EPA's Toxic Release Inventory (TRI) as a carcinogen. It enters the environment during the manufacture, use, and disposal of styrene-based products. The interaction of styrene with soil is a very complex phenomenon and means are required to understand this matter more fully. The objective of this research is to develop a system for the removal and biodegradation of styrene and to determine the mechanisms for styrene removal from soil by biosurfactant (rhamnolipid). The system will consist of two stages: soil flushing followed by an anaerobic biobarrier for groundwater treatment. In this research, a long-term styrene adsorption test was performed. The initial column test results showed that after 90 days, high concentrations (from 700 mg/kg to 4700 mg/kg) remained. A biologically produced surfactant named rhamnolipid (JBR-425) was used in the batch removal tests at a concentration of 0.5%. From the preliminary results, 72.7% of the styrene was removed from the soil after a day and 88.3% after 5 days (original styrene concentration was 32,750 mg/kg). From a 16,340 mg/kg styrene contaminated soil, 90% of the styrene was removed after a day. In the styrene biodegradation tests, the initial results showed that with anaerobic biomass, styrene degraded 37.9% after day 1 and 67% after day 3. Both removal rates were higher than without the biomass (12.5% removal at day 1 and 46% removal at day 3). Also, the rhamnolipid did not inhibit styrene biodegradation. The initial results from the styrene removal and biodegradation tests showed that styrene could be removed from soil by the rhamnolipid and that styrene could be biodegraded under anaerobic conditions with the biomass.

[1]Corresponding author, Tel: 1(514)848-2424, ext. 7925, Fax: 1(514)848-7965, Email:mulligan@civil.concordia.ca

1. Introduction

Styrene has been manufactured in the United States since 1938; it has been used widely in the manufacture of polystyrene plastics, resins and copolymers such as styrene-acrylonitrile (SAN), acrylonitrile-butadiene-styrene (ABS), and styrene-butadiene rubber (SBR). It is also used to make consumer products such as packaging, electrical, and thermal insulation materials, pipes, automotive components, drinking tumblers, other food use utensils, and carpet backing (ATSDR, 1992).

Styrene is classified in EPA's Toxic Release Inventory (TRI) as a carcinogen, it has been stated that releases of more than 0.454 kg of styrene into the air, water, or land must be reported annually and entered into the TRI (NSC, 2004). Styrene enters the environment during the manufacture, use, and disposal of styrene-based products. Mostly, soil and sediments may become contaminated with styrene by chemical spills, landfill disposal of styrene-containing wastes or discharge of styrene-contaminated water (ATSDR, 1992). From the Ministry of Environment Quebec, the Land Protection and Rehabilitation Regulation, the limit of styrene in land is 5 mg/kg soil (in dry matter), which is the maximum acceptable for residential, recreational and institutional sites and commercial sites located in a residential district are also included (Environment Quebec, 2003).

Styrene release to soil is subject to biodegradation. Degradation of 87% to 95% has been observed in the sandy loam and landfill soil during the 16 week incubation. However, styrene is also quite stable in soil. Styrene that leaked into surrounding soil from buried drums persisted in the soil up to two years. It also has been demonstrated that styrene buried in soil can leach into underlying groundwater. It is often found that some portion of the volatile organic compounds (VOCs) may be permanently retained in the soil. The sorbed contaminants are generally thought to be unavailable to participate in environmental fate processes such as biodegradation. The aged compounds that have been exposed to soils for an extended period are found to be less available for desorption or biodegradation than the same freshly added compounds (Liste *et al.*, 2001). Regarding all of these facts, developing a new method for managing styrene-contaminated soil and groundwater is becoming more important.

Surface active agents (surfactants), consist of two portions with opposing characteristics, one part is hydrophilic and the other is hydrophobic (Tsujii, 1998). Biosurfactants are surfactants produced by bacterial and yeasts that are potentially more biodegradable than synthetic ones. By reducing the interfacial tension and forming micelles, many environmental applications demonstrated that surfactants or biosurfactants could improve oil recovery, remove heavy metals from contaminated soil, enhance remediation of hydrophobic organic compounds from soil (Mulligan and Eftekhari, 2003; Wang and Mulligan, 2004; Mulligan, 2005; Urum and Pekdemir, 2004). Some methods focus on the use of surfactants in washing procedures to enhance the remediation of contaminated soil (Mulligan, 2002; Chu, 2003).

Among the available methods and technologies, biotreatment processes are less expensive than other waste treatments and energy requirements for biotreatment are less. Furthermore, treatments can be done onsite, wastes can be degraded completely, and other chemical and physical treatments can enhance the process. Using anaerobic treatment to treat leachates has become a promising technology due to the low quantities of sludge generated from the treatment, methane production and the high quality of the treated effluent (Mulligan,

2002). Since the activity of microorganisms is vital to the process, the success of this treatment is strongly influenced by the concentration of styrene in the influent, which has a major affect on the bacterial population inside the system. In-situ remediation techniques have considerably lower costs and maintenance requirements compared to ex-situ techniques. They also remove the contamination or immobilize the contaminants in-situ without the need to transfer the affected material from the site, thus eliminating the risk of phase transformation of volatile compounds (Yerushalmi *et al.*, 1999). Biobarrier systems, also referred to as permeable bioreactive barriers, are used for the treatment of contaminated soil and groundwater. Due to the limited supply or complete absence of oxygen under in-situ conditions, remediation of leachate or groundwater under anoxic or anaerobic conditions is very important (Yerushalmi *et al.*, 1999).

This research will focus on the use of rhamnolipid for the removal of styrene, which has adsorbed on to soil. Then anaerobic technology will be used to biodegrade styrene in the leachates generated from soil washing (in a batch or continuous process). It is expected to improve the styrene removal technology and develop more understanding in to the removal styrene from soil. Anaerobic fixed bed columns will be used to simulate biobarriers in this study. The system degrades the styrene in the leachate generated by biosurfactant soil flushing. Nutrients are added to enhance the performance.

Therefore, the main objectives of the research are to evaluate the feasibility of using rhamnolipid in styrene removal (using batch and continuous flow tests), to determine the mechanisms for styrene removal from the soil by the rhamnolipid and to develop an in-situ biobarrier using anaerobic fixed bed columns to biodegrade the leachate (contaminated with styrene) from rhamnolipid flushing.

2. Literature Review

In this section, a review will be made of styrene in the environment, its usage and applications. Various technologies have been used on the removal of styrene from water and soil; specifically biotreatment processes will be presented. A brief background on biobarriers, the anaerobic fixed bed reactor, and its applications will be presented. Several types of surfactants and their applications in remediation technology will be also reviewed in this chapter.

2.1. Styrene

Styrene is primarily a synthetic chemical. It is also known as vinylbenzene, ethenyl benzene, cinnamene, or phenylethylene. Styrene is a colorless liquid and has a high volatility (ATSDR, 1995). It dissolves in some liquids, for example alcohol, ether, methanol, acetone and carbon disulfide but doesn't easily dissolve in water. Styrene reacts violently with chlorosulfonic acid, oleum, sulfuring acid and alkali metal graphite. It reacts vigorously with oxidizing materials and readily undergoes polymerization when heated or exposed to light or a peroxide catalyst (IPIS_INCHEM, 1996).

Styrene is normally produced from industrial synthesis, by dehydrogenating ethylbenzene with iron catalyst to produce styrene (Figure 1). Reactions occur at high temperatures (600-

700 °C) (Gibbs and Mulligan, 1997). Since ethylbenzene is a common contaminant, it is also produced by oxidation of ethylbenzene to its peroxide, which then reacts with propylene to produce propylene oxide and alpha-methylphenyl carbinol. The carbinol is then further dehydrated to produce styrene (ATSDR, 1992). Also, the cinnamic acid could make styrene by decarboxylation (Figure 2) (Gibbs and Mulligan, 1997). Low levels of styrene also occur naturally in a variety of foods such as fruits, vegetables, nuts, beverages, and meats (ATSDR, 2002). Table 1 illustrates the properties of styrene.

2.1.1 Applications and Uses

Styrene had been manufactured in the United States since 1938. In the last few decades, it has seen a steady increase in the global production and utilization (Table 2). Its production has increased 1.8% in the decade between 1978 and 1988 and with a 4.8% increase in production during the 5-year period between 1983 and 1988. Production of styrene for 1987 was over 3.6 billion kilograms (ATSDR, 1992). Table 2 listed the distribution of global capacity for styrene production from 1990 to 1992 (Gibbs and Mulligan, 1997). In 2004, the total styrene production in the United States was 5.5 billion kilograms, the production in Canada was 0.9 billion kilograms and in Mexico, it was 150 million kilograms (Chemical Profile, 2004).

$$C_6H_5CH_2CH_3 \longrightarrow C_6H_5CHCH_2+H_2$$

Figure 1. Styrene production (adapted from Huntsman, 2001)

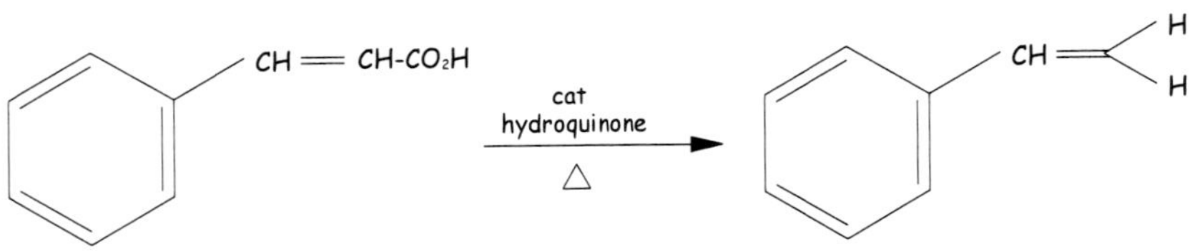

Figure 2. Decarboxylation of cinnamic acid to produce styrene (adapted from Warhurst, 1994)

Styrene has been used predominantly in the manufacture of polystyrene plastics and resins (62%). Some of these resins are used for construction purposes such as in insulation or in the fabrication of fiberglass boats. Styrene is also used as an intermediate, in the synthesis of materials used for ion exchange resins and to produce copolymers such as styrene-acrylonitrile (SAN), acrylonitrile-butadiene-styrene (ABS), and styrene-butadiene rubber (SBR)(ATSDR, 1992). Styrene also been used to make consumer products such as packaging, electrical, and thermal insulation materials, pipes, automotive components, drinking tumblers, other food-use utensils, and carpet backing (ATSDR, 1992).

Table 1. Properties of styrene (EPA-OPPT, 1994)

Trade Names and Synonyms	Vinylbenzene Phenethylene Cinnamene Diarex HF 77 Styrolene Styrol Styropol
Molecular Formula	C_8H_8
Chemical Structure	$C_6H_5\,CH{=}CH_2$
Physical State	Colorless to yellowish oily liquid
Molecular Weight	104.14
Melting Point	-30.6°C
Boiling Point	145-146°C
Water Solubility	310 mg/L at 25°C
Density	0.9059 g/ml at 20°C
K_{OC}*	920
Log Kow*	2.95
Vapor Pressure	573.14 Pa at 15°C 666.45 Pa at 20°C 1266.25 Pa at 30°C 1332.89 Pa at 35°C
Flash Point	31°C
Henry's Law Constant	275 Pa. m^3/mol
Odor Threshold	0.036 mg/m^3

* **K_{OC}:** Soil Organic Carbon-Water Partitioning Coefficient (Yong *et al.*, 1992)
* **Kow**: Octanol - Water partition coefficient (Yong *et al.*, 1992)

Table 2. Distribution of global capacity for styrene production (Gibbs and Mulligan, 1997)

Location	1990 (10^6 kg)	1991 (10^6 kg)	1992 (10^6 kg)
US	4,035	4,610	4,770
Western Europe	4,025	3,922	4,185
Japan	2,043	2,238	2,398
East Asia	1,070	1,475	1,860
Eastern Europe	1,411	1,451	1,451
Canada	795	800	725
Latin America	562	575	575
Middle East	360	369	360
South Asia	165	165	218
Total	14,466	15,596	16,542

2.1.2. Styrene in the Environment

The US EPA offices regulating styrene are the Water Regulations and Standards, Emergency and Remedial Response, Solid Waste, and Toxic Substances. The Food and Drug Administration regulates styrene as a food additive-synthetic flavoring substance, as an indirect food additive, as a component of polymers in paper in contact with dry food, and regarding the residual styrene monomer limit in polystyrene intended for use in contact with food (NSC, 2004).

Styrene enters the environment during the manufacture, use, and disposal of styrene-based products. Most industrial releases are into the atmosphere. From the National Safety Council (USA) 1998, a total of 25,612 tonnes of styrene were released into the environment (Table 3), mostly into the air (NSC, 2004). Compared to 1988 (Table 4), the US increased styrene emissions by 54% in 1998 (NSC, 2004).

According to Environment Canada, regarding on-site releases for the 2002 NPRI Pollutants reported for 2001, there was a total of 1883.038 tonnes of styrene released into the environment. A quantity of 1883.037 tonnes of styrene released into air and 0.001 tonnes of styrene were released to the land. Compared to 2000, the total release increased by 7.8 % (CEPA, 2001).

Table 3. US styrene release in 1998 (NSC, 2004)

	Amount (kg)	Proportion (%)
Total styrene Release	25,612,122	100
Air	24,316.366	94.94
Surface Water	6,102	0.02
Underground injection	230,281	0.90
Land	153,162	0.60
Transfer off-site for disposal	906,210	3.53

Table 4. US styrene released from 1988 to 1998 (NSC, 2004)

Year	Styrene Released (kg)
1988	16,608,595
1995	20,113,675
1996	20,174,676
1997	21,087,616
1998	25,612,122

Styrene Degradation

Aerobic Styrene Biodegradation

The two main pathways for the aerobic degradation of styrene have been determined, the first is an initial oxidation of the vinyl side-chain and the other is a direct attack on the aromatic ring (O'Leary *et al.*, 2002).

- Side-chain oxidation: Styrene → styrene oxide → phenylacetaldehyde (PAAL) → phenylacetic acid (PAA)

- Styrene direct ring cleavage: Styrene→ styrene cis-glycol→3-vinylcatechol

Anaerobic Styrene Biodegradation

Degradation of styrene under anaerobic conditions was observed in an anaerobic consortium enriched with styrene and ferulic acid. The styrene biodegradation under anaerobic conditions is proposed as:

Styrene → 2-phenylethanol→ phenylacetaldehyde (PAAL) → phenylacetic acid (PAA)(O'Leary *et al.*, 2002).

Styrene in the Air

When styrene is released into the atmosphere, it will react rapidly with both hydroxyl radicals and ozone with a combined, calculated half-life between 3.5-9 hours (CEPA, 1993). Styrene is also degraded in the presence of NOx and natural sunlight (EPA OPPT, 1994).

Styrene in Water

The principal sources of styrene releases to water are industrial effluents. Styrene has been detected in effluents from chemical, textile, latex, and coal gasification plants. Styrene was also identified in one of 63 industrial effluents at a concentration of <10 μg/L (ATSDR, 1992).

Volatilization and biodegradation are the dominant transport and transformation processes for styrene in water. Styrene is rapidly lost from surface waters by volatilization, with the half-life ranging from 1 to 60 hours (CEPA, 1993), depending on the depth of the water body and degree of turbulence. On the basis of computer models, the half-life of styrene is 3 days in a pond and 13 days in an oligotrophic lake. Styrene could be biodegraded quite rapidly in water under aerobic conditions. The biodegradation half-life of styrene in water was estimated to be less than 5 days (CEPA, 1993). Very low concentrations of styrene (less than 10 μg/L) were almost completely degraded in 20 minutes in an aerobic biofilm reactor after acclimation, but the chemical was persistent in a methanogenic biofilm column. It was found that the rate of styrene biodegradation in groundwater was slow. The half-life of styrene in groundwater was estimated to be between 6 weeks and 7.5 months (ATSDR, 1992).

Araya *et al.* (1999) used anaerobic treatment for effluents from an industrial polymers synthesis plant. The composition of the wastewater included acrylates, styrene, detergents, with a minor amount of silicates and a significant amount of ferric chloride. An upflow anaerobic sludge blanket (UASB) was used in this study. The COD removal rate fluctuated between 55 and 75% until reaching a constant value of 75%. After 3 months, the methane percentage increased until it reached again a level between 70% and 80%. The alkalinity relationship of volatile fatty acid alkalinity and the total alkalinity was between 0.5 and 0.75. Anaerobic digestion seems to be appropriate for the treatment of the residual waters from a chemical polymer's synthesis plant. After a period of adaptation of the microbial populations, it was able to diminish the COD value of the wastewater by 75%.

Styrene in Soil

Soil and sediments may become contaminated with styrene by chemical spills, landfill disposal of styrene-containing wastes or discharge of styrene-contaminated water (ATSDR, 1992). According to the Ministry of Environment and Sustainable Development of Quebec, Land Protection and Rehabilitation Regulation, the limit of styrene in soil is 5 mg/kg (in dry matter), which is the maximum acceptable limit for residential, recreational and institutional sites and commercial sites located in a residential district (Environment Quebec, 2003).

Styrene spills have occurred quite frequently in the last decade. On January 26, 1992, two barges collided on the intercostals waterway close to Wax Lake in Louisiana. One barge carrying a cargo tank with 340,000 litres of styrene started leaked after the collision. The styrene reached the soil on the south embankment. Some of it also polymerized and formed clumps with silt and other solids, causing it to sink (National Oceanic and Atmospheric Administration, 2002).

On February 12, 1990, a fire broke out at a tire dump in Hagersville, Ontario. The fire burned intensely for a total of seventeen days, consuming a total of 12.6 million tires. The incident created 20,000 m^3 of solid waste and contaminated 4.5 hectares of land with fire water containing 12 to 50 m^3 of liquid residue (benzene, toluene, xylene, styrene, oils, etc.). Between 12,000 and 15,000 liters of oil were estimated to have reached the water (Environment Canada, 1984-1995).

On July 20, 2001, a Canadian National/Illinois Central (CNIC) freight train derailed and spilled a total of 84,721 L of styrene. The response operation was divided into four main scopes of work that included air monitoring, spill area operation, road construction and environment assessment. The case was closed in August 2001 (EPA, 2001).

On March 14, 2005, a styrene release occurred at the Shell Chemicals Scotford site, in Strathcona County, Saskatchewan, Canada from an over-pressurized tank containing by-product from the styrene monomer manufacturing process (styrene residue). This residue was recycled as a fuel for plant boilers to produce steam. Site remediation include soil excavation of the tank containment area and surface water ditches, collection and removal of the surface water and melting snow to prevent potential contamination from the property; remediation of road surfaces and parking lots on the site, power washing of the site building and grass mowing and removal of clippings (Shell Chemicals, 2005).

Styrene released to soil is subject to biodegradation. Degradation of 87-95% has been observed in sandy loam and landfill soil over a 16-week period (Fu, 1992). It has been demonstrated that styrene buried in soil can leach into the underlying ground water. Styrene that leaked into the surrounding soil from buried drums persisted in the soil for up to two years. The persistence of styrene under waterlogged conditions may be of environmental importance because many waste sites are anoxic (Fu and Alexander, 1996).

Styrene in water may also partition to soils and sediments. The extent of adsorption of sparingly water-soluble compounds is often correlated with the organic carbon content of the adsorbent. In surface soils, where the amount of organic carbon will be highest, the movement of styrene will be retarded by adsorption. In deeper subsurface environments where the amount of organic carbon may be low, adsorption may not be as significant. Based on field measurements, the rate of movement of styrene in an aquifer was about 80 times slower than that of the groundwater, which is attributed to adsorption (ATSDR, 1992).

Health Effects

Styrene has one of the highest blood/gas phase solubility ratios of the common industrially produced chemicals (Table 5) which means that most of it is transferred to the blood from the lungs during each respiratory cycle. The rate of its absorption will be greatly increased as the rate and depth of respiration increases (Gibbs and Mulligan, 1997).

Table 5. Blood /gas solubility ratios for selected solvents (Gibbs and Mulligan, 1997)

Gas/solvent	Blood/gas phase solubility ratio
Ethylene	0.14
Toluene	3.8
Benzene	7.2
Chloroform	15.0
Styrene	32.0

For an acute effect, exposure levels of 100 mg/L and above cause irritation of the mucous membranes of the eyes and the upper respiratory tract in man. Epidemiological studies on workers with long-term occupational exposure to styrene have shown an increased prevalence of abnormal EEGs (electroencephalogram) associated with urinary mandelic acid concentrations of 700 mg/L or more; at 1600 mg/L a reduction in psychomotor performance and visuomotor accuracy in psychological tests has been observed (IPCS, 1983). The Occupational Safety and Health Administration (OSHA) regulates styrene on the basis of avoidance of narcosis in the work place. In 1987, the International Agency for Research on Cancer (IARC) in Lyon, France upgraded styrene's classification from a Group 3 "not classifiable" to Group 2B "possibly carcinogenic to humans" (Styrene Forum, 2005).

2.2 Surfactants

Materials exhibiting the characteristic of modifying interfacial interactions by way of enhanced adsorption at interfaces are referred to as surface-active agents or surfactants. Materials that possess chemical groups leading to surface activity are generally referred to as being “amphipathic” or amphiphilic (Figure 3) (Myers, 1992).

The applications of surfactants in science and industry are numerous. They range from primary processes such as the recovery and purification of raw materials in the mining and petroleum industries to enhancing the quality of finished products such as paints, cosmetics, pharmaceuticals and food. Table 6 lists the major applications of surfactants.

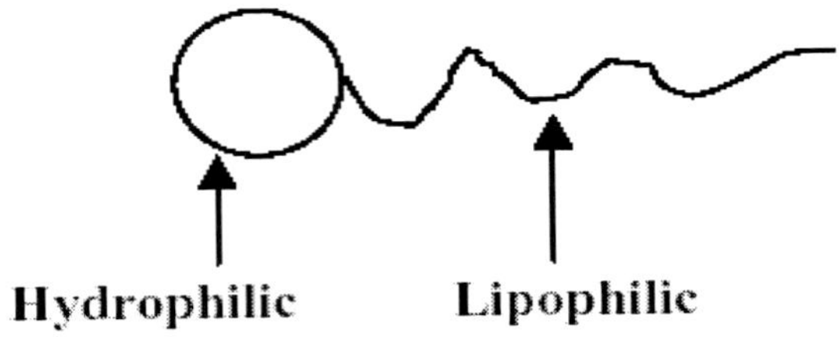

Figure 3. Schematic representation of surfactant molecule

Table 6. Major applications of surfactants (Myers, 1992)

Industrial	Consumer Goods
Agricultural crop applications	Adhesives
Building material	Dry-cleaning fluids
Cement additives	Goods and beverages
Coal fluidization	Household cleaning and laundering
Coating and leveling additives	Pharmaceuticals
Electroplating	Photographic products
Emulsion polymerization	Soaps, shampoos, creams
Industrial cleaning	Water proofing
Leather processing	
Lubrication	
Mold release agents	
Ore flotation	
Paper manufacture	
Petroleum recovery	
Surface preparations	
Textiles	

The most useful chemical classification of a surfactant is based on the nature of the hydrophile, with subgroups being based on the nature of the hydrophobe. Four basic classes of surfactants are presented as (Figure 4):

- Anionic: with the hydrophilic group carrying a negative charge.
- Cationic: with the hydrophile bearing a positive charge.
- Nonionic: where the hydrophile has no charge.
- Amphoteric: which the molecule contains or can potentially contain both a negative and a positive charge.

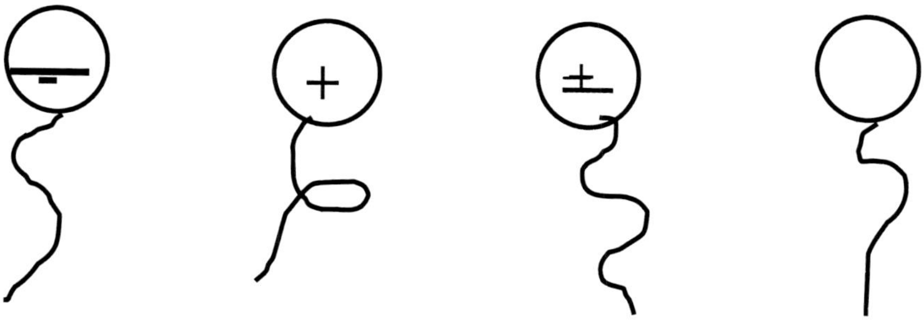

Anionic surfactant	Cationic surfactant	Amphoteric surfactant	Nonionic surfactant

Figure 4. Types of surfactants (adapted from Myers 1992)

The effectiveness of a surfactant is determined by its ability to lower the surface tension, which is a measure of the surface free energy per unit area required to bring a molecule from the bulk phase to the surface. Due to the presence of a surfactant, less work is required to bring a molecule to the surface and the surface tension is reduced (Mulligan, 2005). The surface tension is correlated with the concentration of the surface-active compound until the critical micelle concentration (CMC) is reached. The CMC is defined as the minimum concentration necessary to initiate micelle formation. Figure 5 shows the arrangement of surfactant molecules in the two cases, below and above the CMC. Figure 6 presents how other parameters vary as a function of surfactant concentration.

2.2.1 Biosurfactants

Some surfactants, known as biosurfactants, are biologically produced from yeast or bacteria from various substrates including sugars, oils, alkanes and wastes. Biosurfactants are categorized mainly by their chemical composition and their microbial origin. Biosurfactants are grouped as glycolipids, lipopeptides, phospholipids, fatty acids and neutral lipids. Most of these compounds are either anionic or neutral. Only a few are cationic such as those containing amine groups (Mulligan *et al.*, 2001).

Biosurfactants have several advantages over chemical surfactants, such as lower toxicity, smaller molecular size (less than 1500 Da), higher biodegradability, better environmental compatibility, higher foaming, high selectivity and specific activity at extreme temperatures, pH, and salinity, and the ability to be synthesized from renewable feedstocks. Earlier studies on biosurfactants centered mainly on the properties, biosynthesis, and chemistry and have been reviewed by Wang and Mulligan (2004), Mulligan (2005) and Dahr Azma (2002). Table 7 lists the types and microbial origin of biosurfactants (Mulligan, 2005). However, in the last two decades, significant work on the fermentative production, genetics, and commercial applications of biosurfactants has been performed (Dahr Azma, 2002).

Rhamnolipid

Rhamnolipids, used in this study, are biosurfactants from the glycolipid group produced by *Pseudomonas aeruginosa* (Mulligan, 2005). The surface tension of rhamnolipid of 29 mN/m has been determined. *Pseudomonas aeruginosa* can produce rhamnolipids from substrates including C11 or C12 alkanes, succinate, pyruvate, citrate, fructose, glycerol, olive oil, glucose and mannitol. Composition and yields depend on the fermentor design, pH, nutrient composition, substrate and temperature used (Mulligan, 2005).

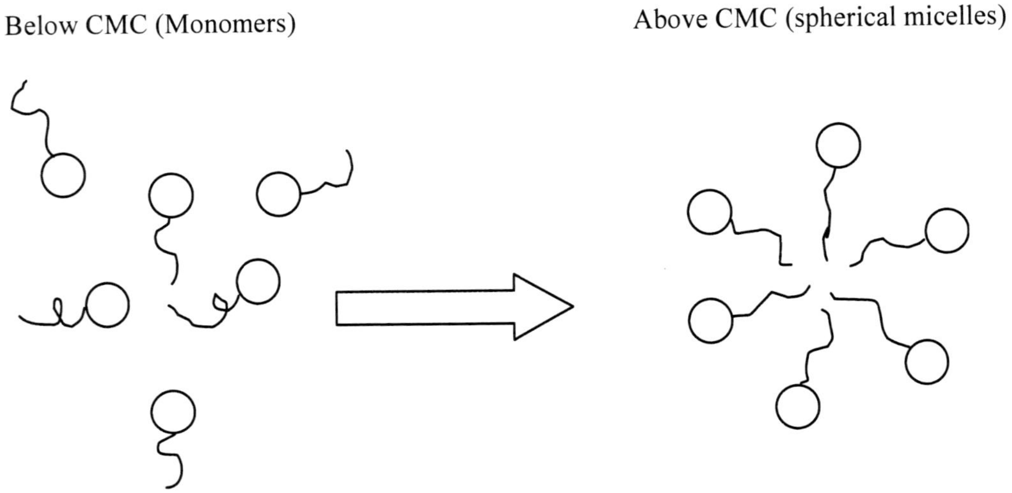

Figure 5. Surfactant micellization (adapted from Becher 1965)

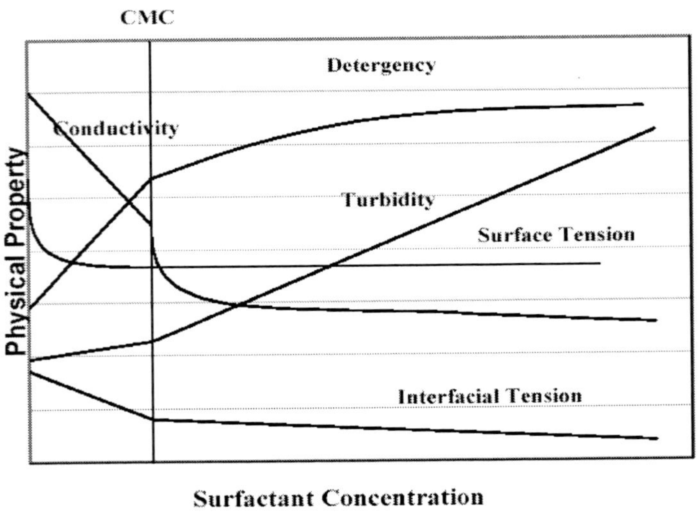

Figure 6. Variation of some important physical properties as a function of surfactant concentration (Myers, 1992)

Table 7. Types and microbial origin of biosurfactants (Mulligan, 2005)

Type of surfactant	*Microorganism*
Trehalose lipids	*Arthrobacter paraffinevs* *Corynebacterium spp.* *Mycobacterium spp.* *Rhodococus erythropolis, Nocardia sp.*
Rhamnolipids	*Pseudomonas aeruginosa* *Pseudomonas sp., Serratia rubidea*
Sophorose lipids	*Candida apicola, Candida bombicola* *Candida bogoriensis* *Candida lipolytica*
Glycolipids	*Alcanivorax borkumensis,* *Arthrobacter sp., Corynebacterium sp.,* *R. erythropolis, Serratia marcescens,* *Tsukamurella sp.*
Cellobiose lipids	*Ustilago maydis*
Polyol lipids	*Rhodotorula glutinus,* *Rhodotorula graminus*
Diglycosyl diglycerides	*Lactobacillus fermentii*
Lipopolysaccharides	*Acinetobacter calcoaceticus (RAG1),* *Pseudomonas sp., Candida lipolytica*
Arthrofactin	*Arthrobacter sp.*
Lichenysin A, Lichenysin B	*Bacillus licheniformis*
Surfactin	*Bacillus subtilis, Bacillus pumilus*
Viscosin	*Pseudomonas fluorescens*
Ornithine, lysine peptides	*Thiobacillus thiooxidans,* *Streptomyces sioyaensis,* *Gluconobacter cerinus*
Phospholipids	*Acinetobacter sp.*
Sulfonylipids	*T. thiooxidans, Corynebacterium* *Alkanolyticum*
Fatty acids (corynomycolic acids, spiculisporic acids, etc.)	*Penicillium spiculisporum,* *Corynebacterium lepus,* *Arthrobacter paraffineus,* *Talaramyces trachyspermus,* *Nocardia erythropolis*
Alasan	*Acinetobacter radioresistens*
Streptofactin	*Streptomyces tendae*
Particulate surfactant (PM)	*Pseudomonas marginalis*
Biosur PM	*Pseudomonas maltophilla*

There are four types of rhamnolipids (Figure 7). Rhamnolipid type I and type II are suitable for soil washing and heavy metal removal while type III is for metal processing, leather processing, lubricants, pulp and paper processing. Type IV is usually used in textiles, cleaners, foods, inks, paints, adhesives, personal care products, agricultural adjuvants, and water treatment (Jeneil, 2002).

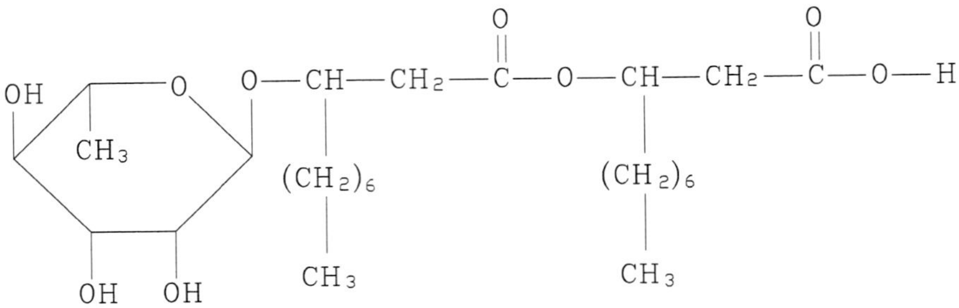

Rhamnolipid Type I

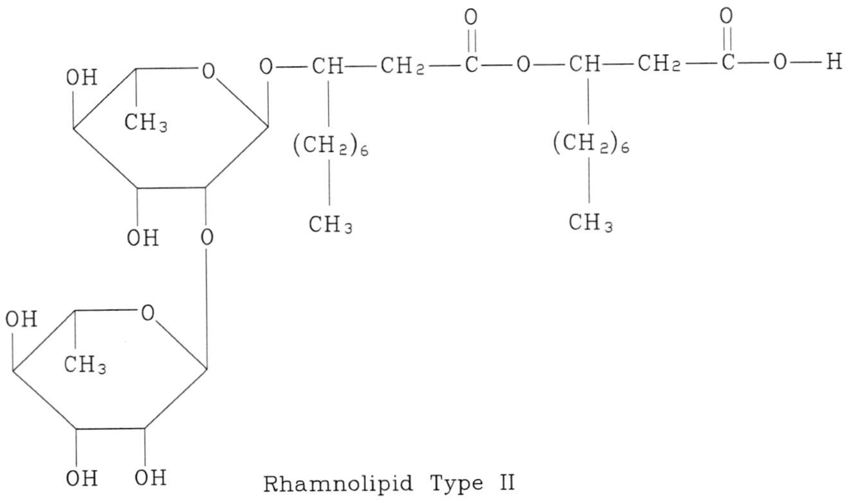

Rhamnolipid Type II

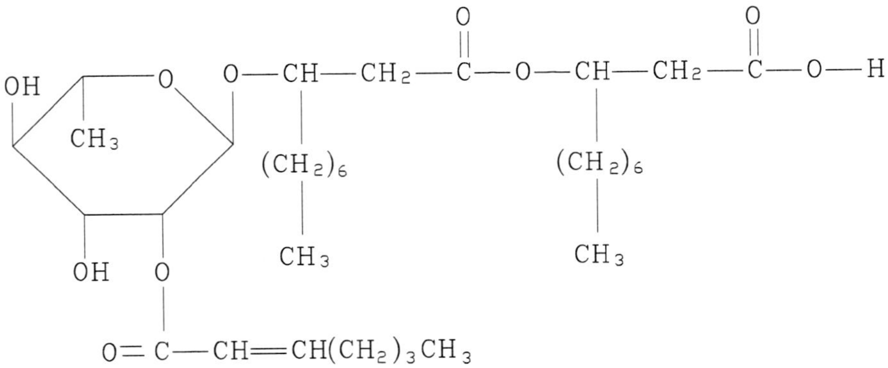

Rhamnolipid Type III

Figure 7. Continued on next page

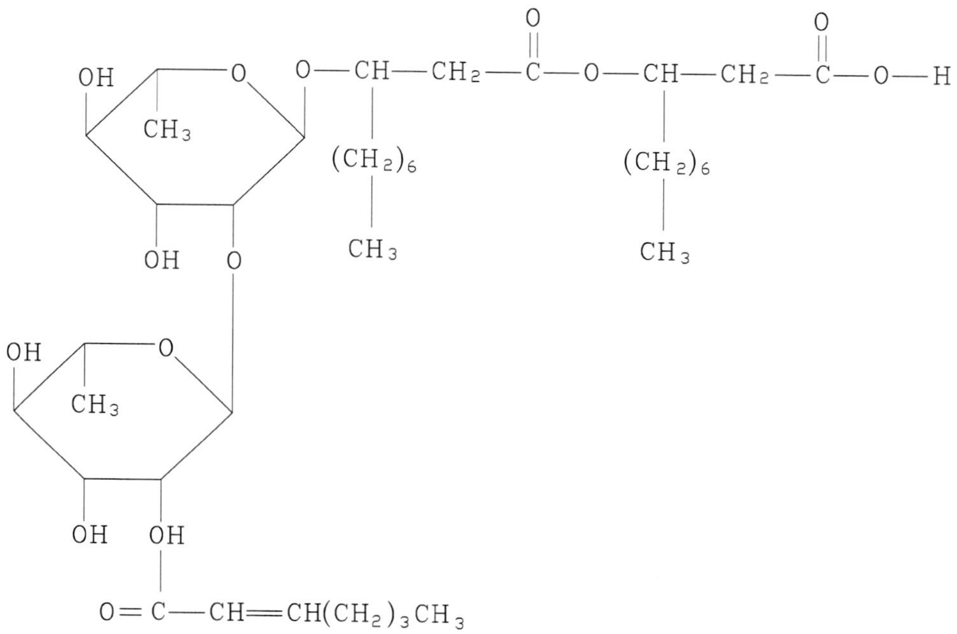

Rhamnolipid Type IV

Figure 7. Different types of rhamnolipids (adapted from Tsujii 1998).

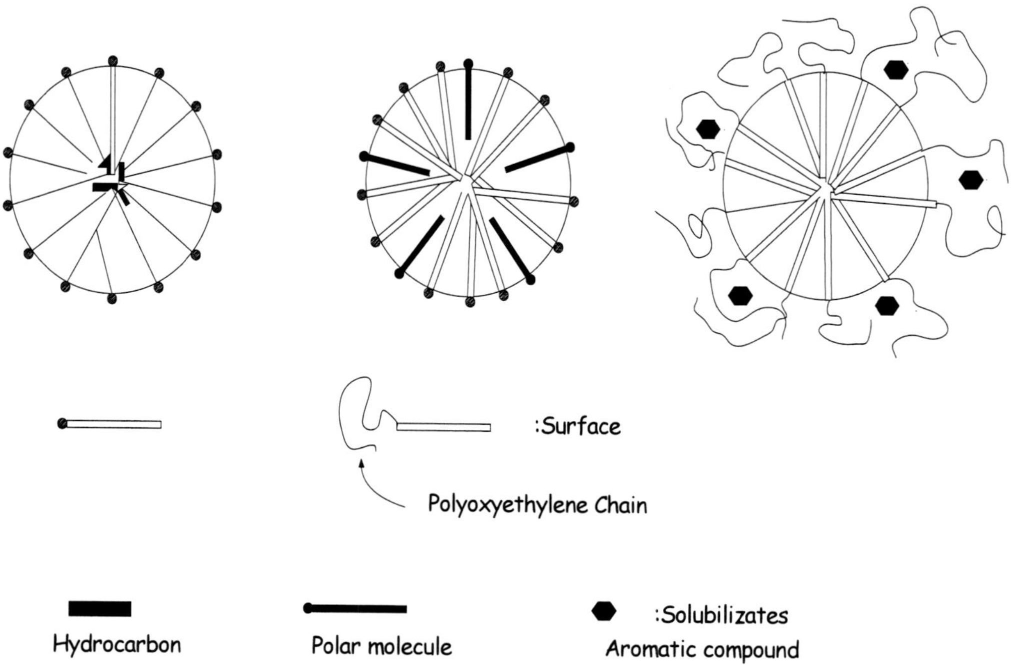

Figure 8. Illustration of the solubilized location of some organic compounds in the micelle (adapted from Tsujii, 1998)

Several studies have examined the effects of rhamnolipid on the biodegradation of organic contaminants; especially focusing on various hydrocarbons with low solubility. It has been shown that rhamnolipid addition can enhance biodegradation of hexadecane, octadecane, n-paraffin and phenanthrene in liquid systems, in addition to hexadecane, tetraadecane, pristine, creosote and hydrocarbon mixtures in soil (Noordman *et al.*, 1998; Mulligan *et al.*, 2001). A couple of possible mechanisms have been postulated to increase the biodegradation. The first is that the substrate solubility increased for the microbial cells and another is the interaction with the cell surface, which increases the hydrophobicity of the surface and allows hydrophobic substrates to associate more easily. It has been determined that rhamnolipid at a concentration of 300 mg/L has increased the mineralization of octadecane to 20% from 5% originally. It also increased the solubility of hexadecane from 1.8 μg/L to 22.8 μg/L (Mulligan, 2005). Urum *et al.* (2003) studied the applications of surfactants in enhancing ex-situ soil washing. Several surfactants including rhamnolipid were used for crude oil-contaminated soil. They determined that rhamnolipid increased crude oil removal to over 80%, and that temperature and surfactant concentration were the most influential parameters. In Figure 8, it is illustrated that different organic compounds were located in the micelle at their most preferential positions (Tsujii, 1998).

Styrene With Surfactant

Doong and Lei (1996) examined the effect of anionic and non-ionic surfactants on sorption and micellar solubilization of monocyclic aromatic compounds including benzene, chlorobenzene and styrene (BCS). Compared to the non-ionic surfactant Triton X-100, the anionic surfactant increased solubilization of aromatic hydrocarbons linearly with increasing concentration. The solubilization of aromatic hydrocarbons was enhanced about 20% to 43% with SDS. Also, surfactant and BCS sorption conformed to the Langmuir isotherm and the adsorption capacity was increased as the hydrophobicity of the organic compounds increased.

2.3. Available Technologies for Removal of Styrene from Soil

2.3.1. Physical and Chemical Treatment

Landfill

A landfill is defined as a system designed and constructed to contain discarded waste to minimize the release of contaminants to the environment (LeGrega and Buckingham, 2001). Isolating contaminated soil and waste in a landfill is the oldest means of reducing exposure. A 8,094 m^2 abandoned landfill located in Harris County, Houston contained over 550,000 kg of liquids and sludges from local chemical industries since 1958. In 1984, the investigation found that the contamination included styrene tars, and their degradation products of substituted benzenes, substituted biphenyls, polynuclear aromatic hydrocarbons, and nonaromatic cyclic hydrocarbons. EPA (2004) decided to excavate the contaminated soils and transport them to a hazardous waste landfill regulated under Subtitle C of the Resource Conservation and Recovery Act.

Incineration

Incineration of solid wastes and contaminated soils is a very affective means of destroying organic wastes. Efficiencies of destruction of much greater than 99% of most organic compounds are routinely observed by maintaining temperatures above 1100°C. Some problems are often cited for incineration such as the products of incomplete combustion, the generation of toxic emissions due to the presence of other contaminants, the residue of non-combustible contaminants and the high cost of treatment (Reible, 1999). In 1990 in Texas City, an on-site incineration was operated with the contaminants including styrene tars, volatile organic compounds (VOCs), polychlorinated biphenyls (PCBs) and metals. The MOTCO site was established in 1959 for the recycling of styrene tars. It was damaged by hurricane Carla in 1961. From 1961 to 1968, the site was used for disposal of chemical waste from local industry. The maximum concentration of styrene was 22,000 mg/kg. The DRE (destruction removal efficiency) by incineration reached the goal of 99.99%. The whole operation lasted for 16 months and cost 110 million US dollars and treated a total of 23,021,000 kg of materials (EPA MOTCO, 2001). The result showed that this technology met the requirements from the EPA.

2.3.2. Bioremediation and Biodegradation

Biotreatment processes are often less expensive and require less energy requirements than other waste treatment processes. Furthermore, the treatment can be done on site, waste can be degraded completely and other chemical and physical treatments can enhance the processes (Mulligan, 2002). Styrene can have a very high biodegradability. Within 16 weeks in a landfill soil and a sandy loam, 87% to 95% of styrene was converted to CO_2, respectively (Fu and Alexander, 1992). Using bioremediation to treat styrene in soil has a lot of potential.

It had been found that using a methanogenic consortium isolated from anaerobic sludge, suggests a possible route of biodegradation of styrene through the addition of water to the unsaturated side-chain; this initially forms 2-phenylethanol and then phenylacetic acid (Grbic-Galic *et al.*, 1990). This compound was readily biodegraded under aerobic conditions (Fu and Alexander, 1992). First-order rate constants for all anaerobic groundwater studies ranged from 0 to 0.016/day. Fu and Alexander (1996) used aquifer material, which had never been exposed to styrene. It was reported that more than 40% of the initial concentration remained after 260 days. No metabolites were detected over this time. The aquifer sediment was collected under aerobic conditions. The loss of styrene in waterlogged soils was also measured. Rapid initial loss was most likely due to the presence of oxygen, followed by a much slower rate of degradation. Nearly 30% of the initial styrene remained after 260 days. It was shown that styrene can be anaerobically biodegraded (Grbic-Galic *et al.*, 1990) which indicates that biodegradation of this compound does not absolutely require the presence of oxygen. However, the persistence of styrene in both anaerobic soils and aquifer sediments suggests that toxic products from the biodegradation of this compound may build up and inhibit further metabolism (Fu and Alexander, 1996).

Biopiles

Biopiles are facilities that use the bioremediation process to economically cleanup hydrocarbon-contaminated soils containing gasoline, diesel, and jet fuels. These systems consist of an aeration system to provide oxygen to the microbes, an irrigation/nutrient injection system to provide nutrients and moisture after pile construction, and a leachate collection system for controlling excess moisture in the pile. A liner, berm, and cover protect the soil piles from storm events and prevent the spread of contaminants (Mulligan, 2002). Remediation time varies depending on hydrocarbon concentrations, soil, nutrients, temperature, and microbial conditions. A bio-pile is a variation of landfarming that can be used if there is not enough space for landfarming. The piles are usually between 100 and 1,500 m^3 and costs were between \$25 and \$85 per 1000 kg for contaminated soil (Mulligan, 2002). Benaxon and Belanger (1994) used a bio-pile to treat styrene-contaminated soil from a polystyrene plant in Canada. After 80 days of treatment, the styrene concentration dropped from 325 mg/kg to 0.01 mg/kg, which is a removal efficiency of 96.9% (Benaxon and Belanger, 1994).

In-situ Biobarrier

In-situ remediation techniques have considerably lower costs and maintenance requirements than ex-situ techniques. They also remove the contamination or immobilize the contaminants in–situ without the need to transfer the affected material from the site, thus eliminating the risk of phase transformation of volatile compounds. Biological in-situ techniques are particularly attractive since they can completely destroy the target contaminants leaving potentially non-toxic chemicals as the products of biodegradation (Yerushalmi *et al.*, 1999).

Biobarriers can be a cost effective technique for remediation of contaminated aquifers. The permeable reactive zones or barriers are installed within aquifers. As contaminated groundwater moves under natural or induced hydraulic gradients through a barrier, the contaminants are degraded (Kao and Lei, 2000).

Yerushalmi *et al.* (1999) studied a gasoline-contaminated water under limited oxygen conditions that was treated in a packed-bed biobarrier system. The gasoline removal efficiencies ranged from 94% to 99.9 % in the protruded stainless steel pieces packed biobarriers and from 86.6% to 99.6% in the granulated peat moss packed biobarrier. Both packed materials in the biobarriers achieved near complete removal efficiencies.

Kao *et al.* (2001) studied using a biobarrier system, which included a peat layer to enhance the anaerobic reductive dechlorination of PCE in-situ. A series of laboratory-scale continuous flow glass columns were used in experiments including three soil columns and a peat column. Anaerobic activated sludge collected from an industrial wastewater treatment plant was used as the microbial inoculum. A flow rate of 0.25 L/day simulated PCE-contaminated water was pumped into the system and up to 98% of PCE removal efficiency was achieved.

An in-situ two-layer biobarrier was used in remediation of PCE-contaminated aquifer (Kao *et al.*, 2003). A laboratory-scale column experiment was conducted to evaluate the feasibility of the system for the bioremediation of PCE-contaminated groundwater. Seven columns were used in series including a soil column, an organic releasing material column, two consecutive soil columns and an oxygen-releasing material column, followed by two other consecutive soil columns. Anaerobic acclimated sludges were inoculated in the first four columns and aerobic acclimated sludges were inoculated in the last three columns to provide microbial consortia for degradation. A flow rate of 0.25 L/day of PCE-contaminated water was pumped as influent, effluent from each column was measured and 99% of PCE-removal rate was achieved (Kao *et al.*, 2003).

Miller *et al.* (2001) studied an oxygenated in-situ biobarrier for MTBE groundwater plume in California. It has demonstrated that MTBE-impacted groundwater can be remediated in-situ through aerobic biodegradation under natural-flow conditions. The biobarrier system has achieved an in-situ treatment efficiency of greater than 99.9 percent for dissolved MTBE and BTEX concentrations. The installation and operation costs were 50 percent lower than a conventional pump and treat system.

Anaerobic Treatment

Anaerobic digestion is a complex series of biological processes that take place in the absence of oxygen and by which organic matter is decomposed and produces a gas principally composed of methane (CH_4) and carbon dioxide (CO_2), otherwise known as biogas. Anaerobic digestion (AD) has been used as the treatment of choice for domestic and industrial wastewater in wastewater treatment plants for many years, due to its advantages such as the production of bioenergy for sustainable development, the conservation of ammonia in the digested material, and the low sludge yield. The process is often a net energy producer, which makes the operating costs significantly lower than those of an aerobic one (Liu *et al.*, 2004). Anaerobic decomposition of complex organic matter into methane and carbon dioxide is a multistep process involving a well-organized community of various microbial species. The main degradation steps of an anaerobic process include hydrolysis, acidogenesis, acetogenesis, and methanogenesis (Liu *et al.*, 2004). Anaerobic digestion reduces the odor and liquid waste disposal problems and produces a biogas fuel that can be used for process heating and/or electricity generation.

The following reaction is typical of anaerobic digestion:

$$\text{Organic matter} + H_2O \rightarrow CH_4 + CO_2 + \text{biomass}$$

DiBacco *et al.* (1997) studied the anaerobic degradation of styrene. Various concentrations of styrene were evaluated. The bacteria were successfully acclimatized to styrene as indicated by the gas production and pH decrease. The surfactant, S-405LF, was used to improve the styrene degradation. 75% of styrene removal was achieved with surfactant and 30% of styrene removal without surfactant.

The interaction of styrene with soil is a very complex phenomenon and styrene-contaminated soil presents a serious environmental problem. It may persist in soil over two years and affect the groundwater. According to the available literature less attention has been paid to removal of styrene from soil. The conventional management of soil containing styrene by landfill disposal and incineration has limitations due to the high operating costs and low styrene removal efficiency. The shortcomings have created a strong incentive for developing an innovative process of combined treatment of styrene-contaminated soil by rhamnolipid washing followed by anaerobic biodegradation.

Biosurfactants have been widely used in oil recovery, degradation of organic compound and many other applications. It has been shown that rhamnolipid addition can enhance biodegradation of hexadecane, octadecane, n-paraffin and phenanthrene in liquid systems, in addition to hexadecane, tetradecane, pristine, creosote and hydrocarbon mixtures in soil (Noordman *et al.*, 1998; Mulligan *et al.*, 2001).

Rhamnolipids are employed to enhance styrene removal and to develop new techniques for styrene removal. In the present study, since very limited studies have been performed to investigate this method, more tests are required regarding the mechanisms of the action. In-situ treatment is preferable to ex-situ washing since the cost and environmental damages due to excavation and transportation of the soil are omitted. Compared to ex-situ techniques, in-situ remediation has lower costs and maintenance and also eliminates the risk of phase transformation. The greatest disadvantage of in-situ bioremediation is the difficulty regarding aeration. Due to the limited supply or complete lack of oxygen under in-situ conditions, the development of an in-situ biobarrier under anoxic or anaerobic conditions becomes very important. In this study, an anaerobic biobarrier will be used to treat leachate from soil washing as shown in Figure 9.

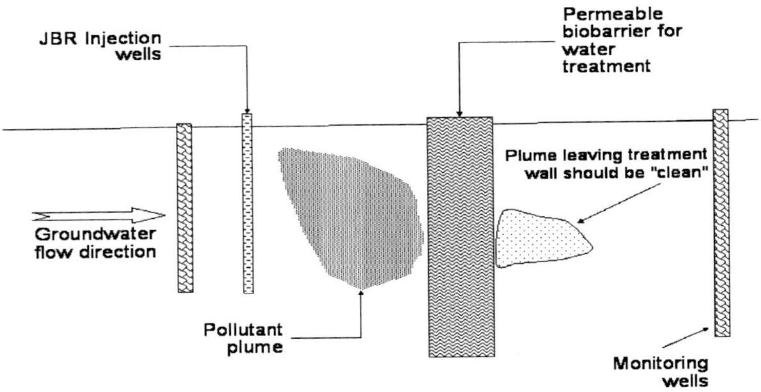

Figure 9. Schematic of proposed in-situ system

3. Materials And Methods

This section is devoted to describing the materials used in this study, the configuration of the experiment and the methods for determine different parameters such as styrene concentration, and chemical oxygen demand (COD). The commonly used materials such as lab-distilled water are not mentioned here.

3.1 Materials

The styrene used in this study was purchased from Fisher Scientific Canada Ltd. Styrene, at 99%, and stabilized with 10-15 mg/L of *p*-tert–butylcatechol was kept in the fridge at 4°C at all times in an amber bottle. Ottawa coarse sand was used in this study. The sand was obtained from construction lab in Concordia University. Before use, the sand has been washed with distilled water then dried at 105°C for 24 hours. Bentonite was collected from the construction lab in Concordia University. Soil used in this study was collected from a construction site (below 30 cm from the soil surface) located on Rue Saint Antoine, Montreal.

Quebec. After collection, all the soil samples were dried at 105°C for 24h and passed through a sieve with a diameter of 2 mm to remove the coarse materials.

Rhamnolipid (JBR425)

The biosurfactant (JBR425) used was a rhamnolipid with the trademark JBR425 from the "JENEIL BIOSURFACTANT CO., LLC" 2002. JBR425 is an aqueous solution of rhamnolipids at 25% concentration. Table 8 lists the physical and chemical properties of JBR425. It is produced from a sterilized and centrifuged fermentation broth that has had most of the protein removed. Two major rhamnolipids, RLL (R1) and RRLL (R2) are present (Jeneil, 2002). Chemically, rhamnolipids are glycosides of rhamnose(6-deoxymannose) and β-hydroxydecanoic acid.

Table 8. Physical and Chemical properties of JBR425 (Jeneil, 2002)

Physical/ Chemical property	Description
Appearance	Dark reddish-brown solution
Concentrations	Viscous suspension
Odor	Soapy
Specific gravity	1.05-1.06
pH	6.5 – 7.5
Solubility in water	Soluble at neutral pH
Suitable diluents	Water and most common alcohols
Surface tension	26 mN/m
Toxicity	68.4% on day 10 of the 28 day test, OECD 301D[†] EC50 > 1000 mg/L, OECD209[†] EC50 of 36.1 mg/L, OECD 202[†]

[†]Toxicity tests reported by Jeneil Biosurfactant (Jeneil Biosurfactant Co. 2002) as accomplished in accordance with Organization for Economic Cooperation and Development (OECD): OECD 301D for ready biodegradability, OECD 209 for activated sludge respiration inhibition (ASRIT), and OECD 202 for aquatic toxicity to daphnia.

Biomass

The anaerobic biomass (granules) used in this study was obtained from an anaerobic reactor that treats cheese whey generated from a cheese plant, Agropur, located in Notre Dame Du Bon Conseil, Quebec, Canada. The biomass was stored in the refrigerator at 4°C until further usage. Before using the biomass in the experiments, the biomass was put in an incubator at 35°C for 2 weeks with simulated wastewater for acclimation. The simulated wastewater was prepared from chemicals of analytical reagent grade from the lab and purchased from Fisher Scientific Ltd. Tables 9, 10 and 11 show the composition of the simulated wastewater.

Table 9. Wastewater components per liter (Alimahmoodi, 2004)

Compound	Amount (g/L)
$NaCl$	0.6
KH_2PO_4	0.44
$MgCl_2$	0.2
$CaCl_2$	0.2

NH_4Cl	1.2
Trace mineral solution	10 ml

Table 10. Trace mineral solution components (Alimahmoodi, 2004)

Minerals	**Amount (g/L)**
$MnSO_4H_2O$	0.1
$CoCl_2$	0.0131
$CaCl_2$	0.076
$CuCl_2$	0.02
$ZnCl_2$	0.1
NaCl	1.0
$NiCl_2.6H_2O$	0.12
$FeCl_3.6H_2O$	1.34

Table 11. Peptone and yeast extract added in solution to the substrate (Alimahmoodi, 2004)

Solution	**Amount**
Sodium acetate	1.5 g/L
Acetic acid	0.75 ml/L
Yeast extract	0.8 g/L
Peptone	0.4 g/L

3.2 Analyses

3.2.1 Organic Matter Determination

According to the ASTM method (1998), ASTM D 422-63, the soil total organic matter (OM) was measured by a dry combustion test (Valdosta State University, 1999) as follows. Five grams of oven-dried soil were weighed and placed in the beaker. The beaker was placed into the furnace for 550°C for 2 hours. After two hours, the furnace was turned off, and allowed to cool overnight. The samples were re-weighed. The weight of the organic matter combusted was determined and then the percentage of organic matter (OM) was determined as:

$$\%\ \text{Organic Matter} = \frac{\text{Total weight before test - Total weight after test}}{\text{Total weight before test}} \times 100\%$$

The soil organic matter content was determined to be 5.3 %.

3.2.2 CMC Measurement of Rhamnolipid

Prior to deciding on the biosurfactant concentration, it is vital to identify the critical micelle concentration (CMC) of the biosurfactant. By increasing the biosurfactant concentration, the surface tension decreases until it reaches the CMC. The CMC refers to the minimum concentration of the biosurfactant required for the formation of micelles. For optimal performance, the concentration of the biosurfactant needs to be of at least equal to its CMC value. In this experiment, the CMC value of the biosurfactant was determined indirectly using surface tension measurements of solutions of various dilutions. First, thirteen concentrations of JBR-425 (0.0001%, 0.0005%, 0.001%, 0.005%, 0.01%, 0.03%, 0.05%, 0.1%, 0.5%, 1%, 2%, 3%, and 4%) were prepared. Then the surface tension was measured using a FISHER SCIENTIFIC Surface Tensiomat (Model 21). Subsequently the surface tension measurements were plotted against the rhamnolipid concentration using a semi log curve. The CMC value for rhamnolipids was found to be 0.03 g/L (Figure 10). The result was close to the previously reported rhamnolipid CMC value by Zhang and Miller (1992) and Wang and Mulligan (2004). During the preliminary tests, a solution of 0.5% of rhamnolipid was prepared by adding 2 ml biosurfactant to 100 ml water.

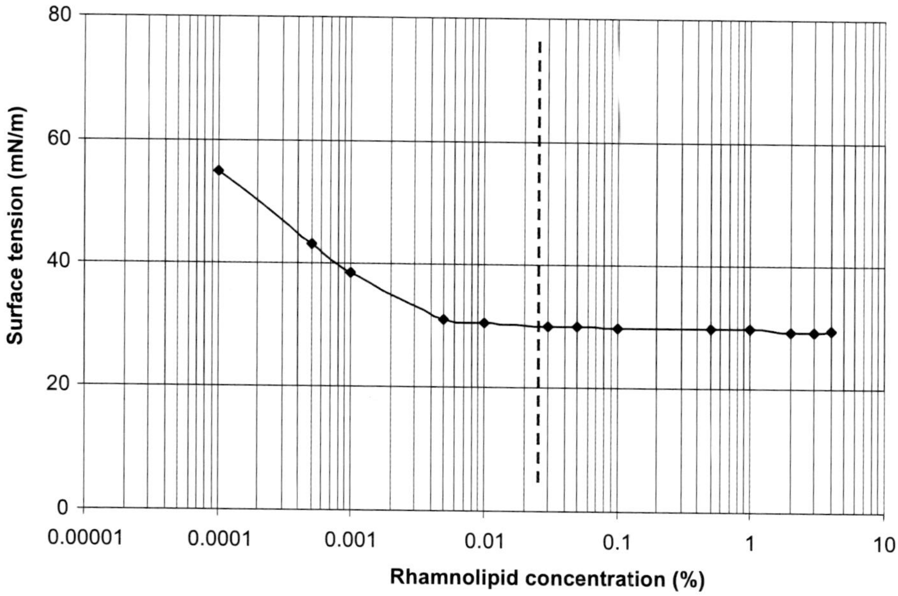

Figure 10. CMC determination by surface tension measurement

3.2.3. Chemical Oxygen Demand (COD)

The Chemical Oxygen Demand test measures the chemical oxidant required to break down organics. COD is an indicator of the concentration of organics in the solution. The Closed Reflux, Colorimetric Method from American Public Health Association (1995) was used. 2.5 ml of sample was added to the standard tube which contains 7.5 ml of the digestion solution (commercially available pre-measured solution containing a 7.5 ml mixture of sulfuric acid, potassium dichromate, silver sulfate, mercuric sulfate, and sulfuric acid in twist cap digestion vials, and potassium hydrogen phthalate (KHP) standard (425 mg KHP/L distilled water; having a theoretical COD of 500 mg O_2/L). the samples were digested for two hours at 150°C and then cooled. The test color was measured using a Perkin-Elmer Lambda Spectrophotometer.

3.2.4. Instrumentation

A UV/VIS Spectrophotometer (Perkin Elmer Model Lambda 400) was used to measure the styrene concentration and the chemical oxygen demand (COD). Styrene was measured at 246 nm. Chemical oxygen demand (COD) was determined at 600 nm.

3.3 Experimental Set Up

Since styrene adsorbs to rubber or polystyrene material, and can degrade under light, amber vials were used in the experiments. Thus, non-glass material was avoided in this study.

3.3.1 Soil Adsorption Tests

Batch Adsorption Tests

For this batch test, six different styrene concentrations were chosen (12 mg/L, 40 mg/L, 60 mg/L, 80 mg/L, 120 mg/L, 200 mg/L) and in triplicate. 1.5 g of the dry soil was transferred to a glass centrifuge tube, and 15 ml solution was added to a glass centrifuge tube. The glass centrifuge tubes were covered with aluminum foil to prevent light exposure. The tubes were put in a container and on a Thermolyne/Barnstead adjustable orbital shaker model M66025 that was set at 105 rpm for 24 hours. After 24 h, all the centrifuge tubes were put in a centrifuge IEC HN-SII for 20 min at 2500 rpm. The supernatant was filtered with a 0.45 μm filter and the concentration of the supernatant was measured with the UV spectrophotometer. Styrene solutions were run as controls to account for the styrene loss from the degradation process. Soils with distilled water were also run as controls to account for any interference due to soil components at 246 nm. To determine how styrene adsorbed to different materials, batch tests were also done with Ottawa sand and bentonite using the same procedure in triplicate.

Long-Term Styrene Adsorption Test

For the long-term test, a glass column (30 cm long and 5.3 cm ID) was filled with soil and styrene was added for a certain period to simulate how styrene would exist in a deep soil, which is close to anaerobic conditions.

1.Test Column 1

Two glass columns were used. Each was filled with 350 g soil and 175 g of sand on the top (Figure 11). 150 ml of 120 mg/L styrene was added, the whole column was covered with aluminum foil after both ends were sealed. The columns were put in the fume hood for 60 days. Next, 1.25 g of sample (since each gram of soil has 0.28ml liquid) was taken from each 3 cm of the column. Methanol was used to extract styrene from the soil. 10 ml of methanol was added and shaken at 105 rpm for 15 min. The supernatant was filtered with a 0.45 μm filter and the concentration of styrene was measured with the UV spectrophotometer at 246 nm. This test studied different adsorbabilities of sand and soil with styrene at a low concentration (120 mg/L) for a long contact time (60 days).

2.Test Column 2

For the second test column, two glass columns were filled with 450 g of soil. 12.5 ml of pure styrene and 100 ml of distilled water (equal to a concentration of 113.75 g/L) were added. The columns were covered with aluminum foil after both ends were sealed with rubber stoppers. The columns were put in the fume hood for 90 days. Next, 1.2 g of the soil sample were taken from each 2 cm of the column, 10 ml of methanol were added and the mixture was shaken at 105 rpm for 15 min. The supernatant was filtered with a 0.45 μm filter and the concentration of styrene was measured with the UV spectrophotometer at 246 nm.

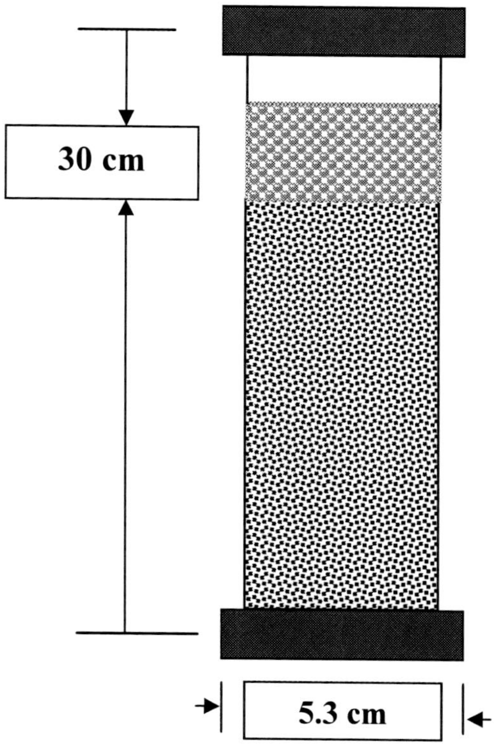

Figure 11. Schematic representation of 175 g of sand and 350 g of soil in test column 1

3.3.2. Styrene Removal Test

Batch Removal Test

Two concentrations were chosen from the long-term column tests for the batch removal test. The first soil sample was taken from the 4-6 cm zone in the column and the second soil sample was taken from the 6-8 cm zone from the column. Two grams of soil were transferred into glass centrifuge tube and 10 ml of 0.5% rhamnolipid was added in each glass centrifuge tube and the following procedure was used. All glass centrifuge tubes were covered with aluminum foil. All tubes were placed on a Thermolyne/Barnstead adjustable orbital shaker model M66025 that was set at 105 rpm for 24 hours. All samples were centrifuged at 2500 rpm for 20 min. The supernatant was transferred to the amber vials. 10 ml of methanol were added to the glass centrifuge tube and was shaken for 20 min at 105 rpm on the shaker. The glass tubes were centrifuged and then the supernatants were filtered with a 0.45 μm-filter. The styrene concentrations were determined in the supernatants by the UV spectrophotometer at 246 nm.

3.4. Styrene Biodegradation Test

3.4.1 Biodegradation of Styrene with Biomass

For the biodegradation test, a 285 mg/L styrene solution was prepared. 10 ml of styrene solution alone or 10ml of styrene with 1ml of biomass were placed in the amber vials and put in the incubator at 35°C. All vials were covered with screw caps to make sure no oxygen transferred so as to simulate anoxic or anaerobic conditions. The time study was chosen as 1 day and 5 days. Next, biomass was washed by the distilled water. Then 10ml of methanol was added to extract styrene from the biomass.

3.4.2 Biodegradation of Styrene in Biosurfactant with Biomass

For this test, the supernatants from batch removal tests were filtered by a 0.45 μm filter and then the UV-spectrophotometer was used to measure the concentration designated as C_0. One ml of biomass was added in the amber vial and put in the incubator at 35°C for 24 hours and 72 hours. To prevent styrene volatilization, no nitrogen gas was sparged. The whole system was close to anoxic conditions since no oxygen transferred. After the incubation, the supernatant was filtered with a 0.45μm filter and then the UV-spectrophotometer was used to measure the concentration (C_f). The difference from C_0-C_f was the removal of styrene by the biomass.

4. Results

4.1. Styrene Adsorption Test

Soil, Ottawa sand and bentonite were introduced in styrene adsorption test as sorbents. Different concentrations of styrene were used in this batch test as the sorbate. Several mathematical models have been developed to quantitatively express the relationship between the extent of sorption and the residual solute concentration. The most widely used are the Freundlich adsorption isotherm model and the Langmuir isotherm model.

4.1.1 Batch Adsorption Test

Freundlich Adsorption Isotherm Model (Yong and Mulligan, 2004)

The Freundlich isotherm, which is an empirical formulation, is expressed as:

$X/m = KC_f^{1/n}$

Where,

X/m: amount of styrene adsorbed per unit weight of the soil (mg/g).

C_f: styrene equilibrium concentration

K and n: experimental constants. K is an indication of the adsorption capacity of the soil; n indicates the effect of concentration on the adsorption capacity and represents adsorption intensity. If data conforms to the Freundlich model, a plot of log C versus log (x/m) should give a straight line.

Langmuir Adsorption Isotherm (Yong and Mulligan, 2004)

The Langmuir adsorption isotherm provides a simple mechanistic picture of the adsorption process and gives rise to a relatively simple mathematical expression. The Langmuir isotherm was constructed by plotting $C_f/(x/m)$ against C_f. If the data lies on a straight line then the Langmuir model can be considered to be appropriate. The model could be presented as $C_f/(x/m) = 1/bq_{max} + C_f/q_{max}$

Where:

x/m: amount of styrene adsorbed per unit weight of soil, (mg/g)
C_f: styrene equilibrium concentration.
q_{max}: maximum uptake capacity of the soil, (mg/g)
b: adsorption equilibrium constant ($K_{adsorption}/K_{desorption}$)

Two isotherms were compared regarding styrene contact with soil; the Langmuir isotherm (R^2=0.9127) (Figure 13) fitted more closely than the Freundlich isotherm (Figure 12) (R^2=0.7619).

The Langmuir equation for soil was:

$$C_f/(x/m)=8.3587+C_f/0.484$$

Where b=0.247 L/g, q_{max}=0.484mg/g

For sand and bentonite styrene adsorption tests, four styrene concentrations were used in the standard batch tests, 12 mg/L, 60 mg/L, 120 mg/L, and 200 mg/L. The Ottawa sand was used as the sorbent in this study, the standard ratio of sorbent: sorbate =10:1 according to ASTM 98 D5285 92. The data was then fit to the Freundlich (Figure 12) and Langmuir isotherms and compared. Although the fit was poor in both cases, the Langmuir (Figure 14) fit slightly better. It can be concluded, however, that the Ottawa sand presented a low adsorbability for styrene.

Since remediation of clay is more difficult, clay sorption was examined. For the styrene and bentonite batch test, since for the 10:1 ratio there was no styrene in the supernatant, the 20:1 ratio was chosen which means that 1g of bentonite was added with 20 ml of styrene solution. The result was different from the soil and Ottawa sand. The clay adsorption isotherm fit the Freundlich isotherm (Figure 12) (R^2=0.9321), much better than for the Langmuir isotherm (R^2=0.7794)(Figure 13).

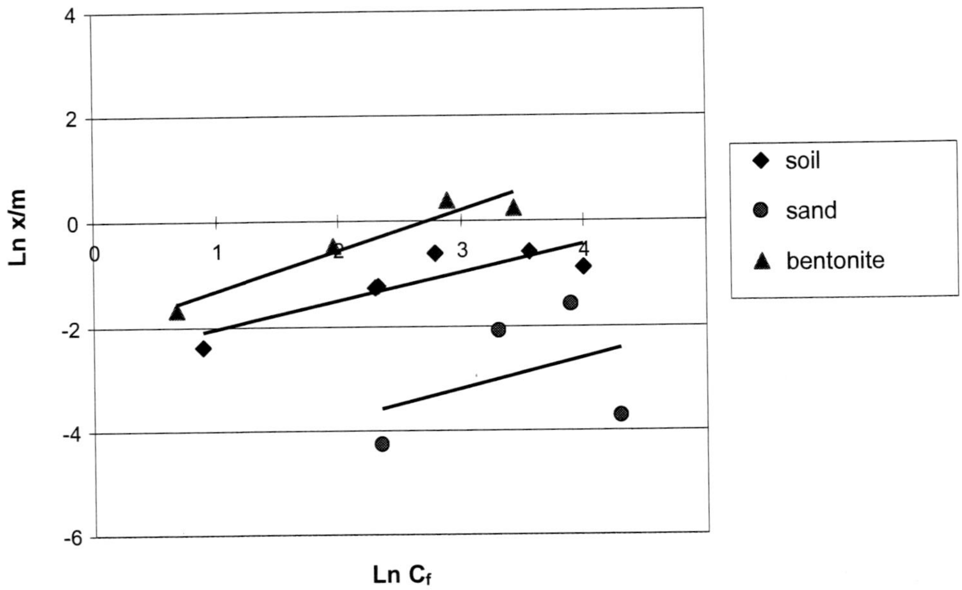

Figure 12. Freundlich isotherm for styrene adsorption on soil, sand and bentonite

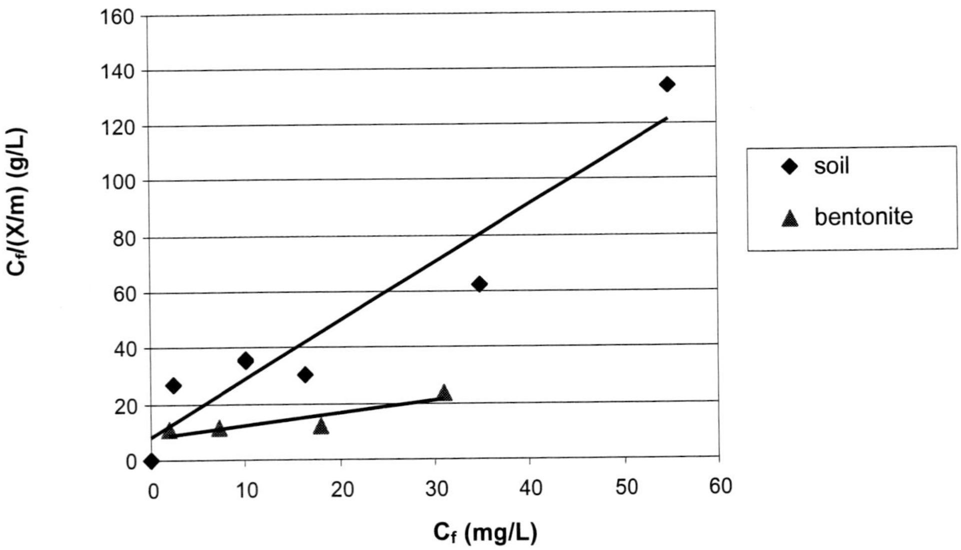

Figure 13. Langmuir isotherm for styrene adsorption on soil and bentonite

The Freundlich equation for bentonite was $x/m=0.123C_f^{1.30}$.

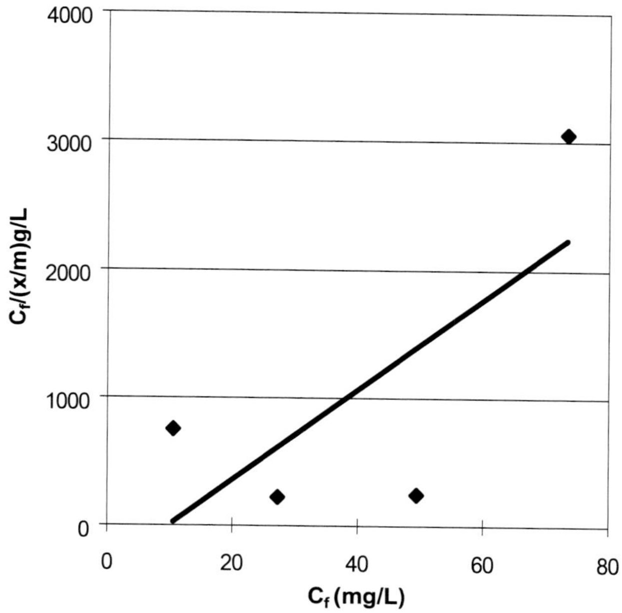

Figure 14. Langmuir isotherm for styrene adsorption on sand

4.1.2. Column Adsorption Test

Test Column 1

In this test, all the soil materials and sand were placed in the glass column and both ends were covered with rubber stoppers. This test studied different adsorbabilities of the sand and soil with styrene at low concentrations (120 mg/L) for a long contact time (60 days).

This column was used to simulate anaerobic conditions and determine the retention or movement of styrene in the column. The concentration of styrene was subsequently determined at different depths after 60 days. Figure 15 shows that compared with soil, sand has low styrene adsorbability. Styrene concentration in the sand layer ranged from 0.25 mg/kg to 0.94 mg/kg. The results also showed that most of styrene remained in the soil layer evenly and styrene concentration in the soil layer ranged from 13 mg/kg to 19 mg/kg, the results showed that styrene (even at low concentration) persisted in the soil system after 60 days.

Test Column 2

In test column 2, the whole column was filled with soil and pure styrene was added. It was also sealed both ends with rubber stoppers and covered with aluminum foil to prevent oxygen transfer and light effects. From Figure 20, the results show that most of the styrene remained in the 4-8 cm zone, at a high concentration from 700 mg/kg to 4700 mg/kg. The lowest concentrations were found in the 16-20 cm zone. The column had been set for 90 days in fume hood. Maximum styrene concentration appeared at the 6 cm zone. It is indicated that in the environment, under anoxic or anaerobic conditions, highly styrene contaminated soil may exist for quite a long time. Styrene may also migrate and reach the groundwater within a certain period.

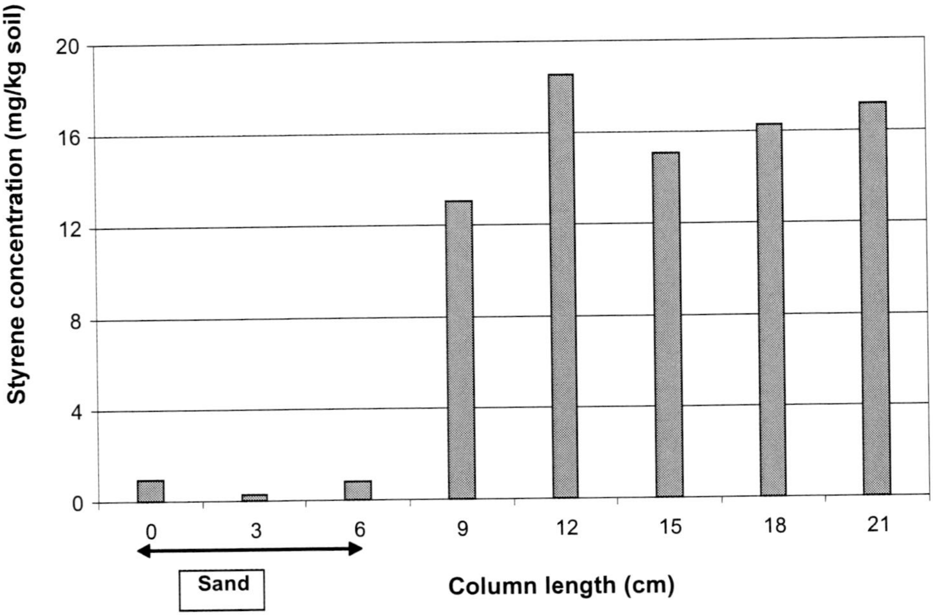

Figure 15. Styrene concentrations in test column 1 after 60 days

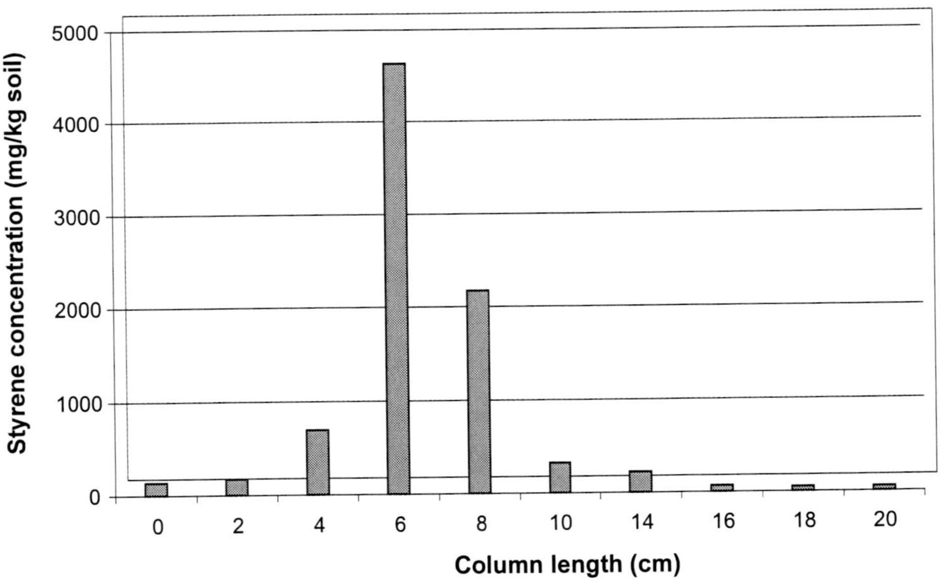

Figure 16. Styrene concentrations in test column 2 after 90 days

4.2 Rhamnolipid (JBR425) Batch Test Washing Styrene

Batch Removal Test

Batch removal tests were carried out where 0.5% rhamnolipid was added as a washing agent to styrene-contaminated soil obtained from test column 2. The original styrene concentration was 32,750 mg/kg soil. From the batch styrene removal test, a longer contact time (5 days) with 0.5% rhamnolipid showed a higher removal rate (Table 12). 72.7% of styrene had been removed on day 1 and 88.3% of styrene had been removed on day 5. To achieve a higher removal rate, a longer contact time would be required.

Another lower level contaminated styrene soil (16,340 mg/kg soil) was used in the batch removal test. On day 1, the removal rate was 90% (Figure 17). The results have indicated the availability of using rhamnolipid as a washing agent to removal styrene from soil. Two concentrations of styrene-contaminated soil (16,340 mg/kg and 32,750 mg/kg) used in the removal test on day 1 achieved a very high removal rate (72.7% and 90%). A higher removal rate was achieved from the lower styrene concentration.

4.3. Styrene Biodegradation Test

Biodegradation of Styrene with Biomass

A test was performed to study the use of biomass to improve styrene biodegradation under anoxic or anaerobic conditions. A styrene solution of 285 mg/L was prepared. Its COD was determined to be 451.7 mg/L. Styrene with biomass was placed in the incubator at 35°C with sealed caps to prevent oxygen transfer. COD and styrene concentrations were determined for day 1 and day 5.

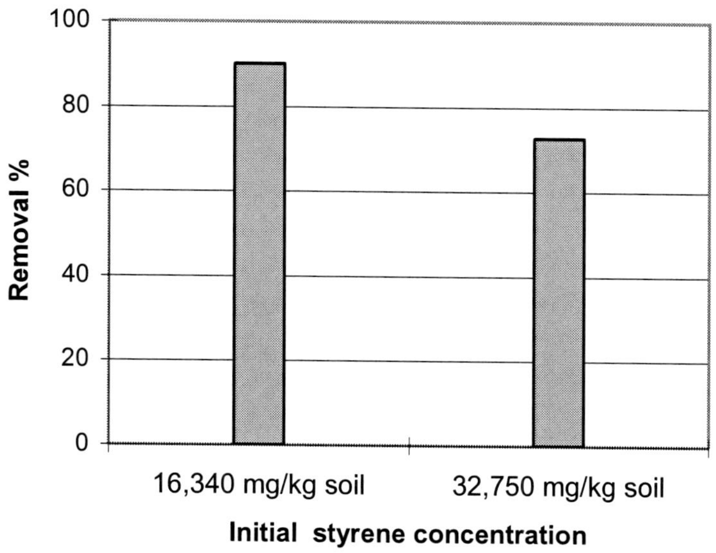

Figure 17. Styrene removal regarding two different concentrations

According to Table 13, the styrene biodegradation test shows that with biomass, on day-1, 37.9% of styrene was degraded and only 12.5% degraded without using biomass. On day-5, total styrene removal rate achieved 67% by using the biomass. It was determined that the biomass could degrade styrene. It was also determined that without using the biomass, styrene could be degraded by 46% after 5 days, which was also a good result. This maybe because the temperature of the incubator (35°C) promoted styrene self-degradation. Therefore, temperature is an important parameter in degradation.

At the end of the experiment, the biomass was washed with distilled water and then 10 ml of methanol was added to extract styrene from the biomass. It was determined that no styrene was found in the extract. Therefore, the biomass did not adsorb any styrene. The removed styrene was degraded by the biomass.

Biodegradation of Styrene in Rhamnolipid with Biomass

After the rhamnolipid batch removal test, the supernatants were transferred to amber vials. This test was carried out to examine if biomass could biodegrade styrene in rhamnolipid environment. Since rhamnolipid is an environmental friendly product, it was shown that rhamnolipid did not inhibit biomass while styrene biodegradation occurred under anaerobic conditions. The initial styrene concentration was 668.5 mg/L. There was a 63.9% styrene removal by the biomass in rhamnolipid on day 1 and 52.1 % of styrene had been removed from the original styrene rhamnolipid supernatants (Table 14). On day 3, 69.1% of styrene was removed with biomass in rhamnolipid and 21.2% of styrene had been removed from the supernatants (Figure 18).

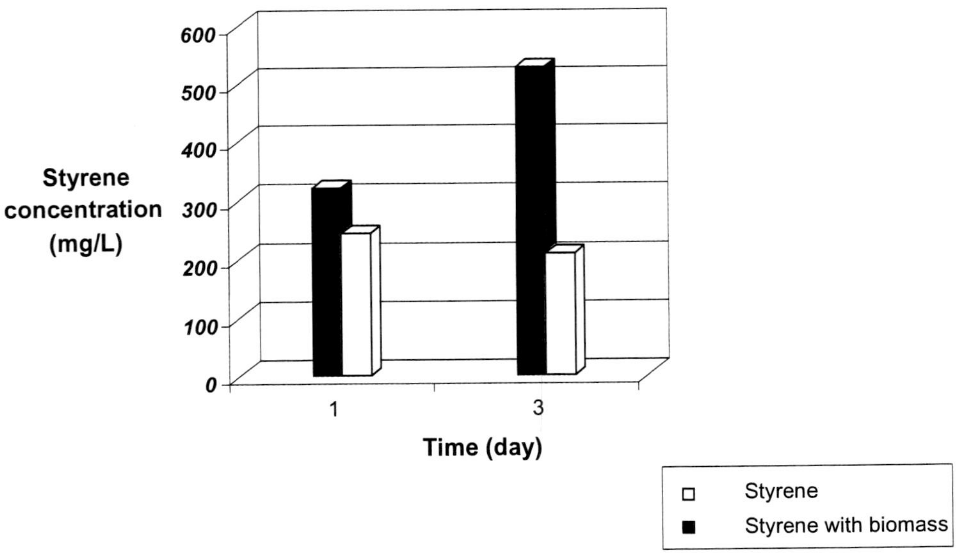

Figure 18. Styrene concentrations with and without biomass after day 1 and day 3 (original styrene concentration of 668.5 mg/L)

On day 1, the biomass only improved the removal by 10% more than without the biomass. On day 3, the styrene with biomass had a much higher removal without biomass.

The styrene concentration of No-biomass test on day 3 was higher than day 1. This result maybe because the rhamnolipid biodegradation inhibited the styrene self-degradation rate or some desorption of styrene from biomass according to a 3 day contact time. Also, some of the styrene biodegradation product may inhibit the further styrene degradation. The styrene in rhamnolipid with biomass showed better removal than without biomass. This means that after the rhamnolipid washing step of the styrene-contaminated soil, the contaminated water that has been generated could be treated with the biomass. Since the rhamnolipid is a biodegradable biosurfactant, the biomass may not be inhibited by rhamnolipid. Since this experiment was carried out for only three days, the data are not sufficient to indicate the styrene removal rate with biomass and rhamnolipid. Further research to determine if styrene in rhamnolipid could be biodegraded with biomass is required.

Table 12. Variation of styrene removal from soil by 0.5 % rhamnolipid

	Initial styrene (mg/ kg)	Styrene remaining in soil (mg/kg soil)	Styrene removal (%)
Day 1	32,750	8,940	72.7
Day 5	32,750	3,840	88.3

Table 13. Styrene and COD removal for an initial styrene concentration of 285 mg/L

Conditions	Styrene (mg/L)	Styrene removal (%)	COD(mg/L)	COD removal (%)
Day 1				
Styrene	249.4	12.5	408.9	9.5
Styrene with biomass	177.1	37.9	239.6	46.9
Day 5				
Styrene	152.9	46.4	240.9	46.7
Styrene with biomass	94.2	66.9	241.7	46.5

Table 14. Styrene removal with and without the biomass (day 1 and day 3). Initial styrene concentration of 668.5 mg/L

Experimental conditions	Styrene (mg/L)	Styrene removal (%)
Day 1		
Styrene	320.5	52.1
Styrene with biomass	241.1	63.9
Day 3		
Styrene	526.6	21.2
Styrene with biomass	206.6	69.1

5. Conclusion

This study demonstrates that rhamnolipid can be used successfully in the remediation of styrene contaminated soil and biomass can be used to biodegrade styrene in anaerobic conditions. Although this study needs further research on biomass with rhamnolipid and the study was investigated in idealized laboratory conditions, the results from this study illustrate some general considerations that are important for the use of rhamnolipid washing styrene contaminated soil.

From the batch adsorption tests, it was shown that compared to sand, clay had a higher styrene adsorption ability. It also showed that clay plays an important role for styrene sorption in contamination site. The Freundlich isotherm for styrene adsorption on bentonite better described the sorption (R^2=0.93).

In the long-term column tests, a styrene concentration of 32,750-mg/kg was found after 90 days under anaerobic conditions. It showed that unlike surface soil, styrene could exist in the subsurface, and the rates of biodegradation and vaporization of styrene were not significant.

In this study, it was shown that it was feasible to use rhamnolipid as a washing agent to remove styrene, which has been sorbed on to the soil. The results show that more than 70% of removal could be achieved for 32,750 mg/kg of styrene after one day and 88.7% removal after five days while a 90% removal was obtained for 16,340 mg/kg of styrene after one day. In the preliminary tests, time was an important factor. Longer contact times (5 days) could achieve higher removal efficiencies (90%).

Anaerobic biomass could biodegrade the styrene and styrene in rhamnolipid under anaerobic conditions. 69.1% of styrene was removed with the biomass solution. The temperature and contact time may affect the biodegradation rate. No biogas was collected and measured, and no styrene metabolite products were analyzed at this time. However, further tests and more detailed analysis will be needed to optimize the process and determine the mechanisms.

Therefore, the combined treatment of styrene-contaminated soil by rhamnolipid washing followed by anaerobic biodegradation process could be a promising technology for in-situ treatment. Compared to other ex-situ treatment, this combined process could be done in-situ without transporting the waste. Compared with landfill or incineration technology, the biotreatment processes are often less expensive and require less energy requirements than other waste treatment processes. Furthermore, the treatment can be done on site and waste could be degraded completely. Using anaerobic treatment to treat leachates has become a promising technology due to lack of oxygen requirements and the high quality of the treated effluent. Rhamnolipid used in this study is an environmental friendly washing agent. After soil flushing, it would not leave any further toxic components in the soil system.

Acknowledgements

The authors would like to thank Jeneil Biosurfactant for supplying the JBR product and Concordia University and NSERC for funding of the project.

REFERENCES

Alimahmoodi, M. (2004). "CO_2 capture and bioconversion to biogas in an anaerobic system using an UASB reactor" {*M. A. Sc. Thesis,* Concordia University, Montreal Canada.

Araya, P., Aroca, G., and Chamy, R.(1999). "Anaerobic treatment of effluents from an industrial polymers synthesis plant". *Waste Management.* 19, 141-146.

ASTDR Fact sheet -Styrene (2002) *http://www.envirotools.org/factsheets/contaminants/styrene.shtml Accessed date 2005.05.22*

ASTDR Toxicological Profile for-Styrene "Production, Import, Use and Disposal" 1992.09 *http://www.atsdr.cdc.gov/toxprofiles/tp53.htm* Accessed date 2005.05.29

Becher, P. (1965). "*In Emulsions, Theory and Practice*" 2[ed] edition, Reinhold Publishing, New York.

Benazon, N., Belanger, D., and Scheurlen, D. (1994). "Bioremediation of ethylbenzene and styrene contaminated soil using biopiles" *Proceedings: 4th Annual Symposium on Groundwaterand Soil Remediation.* September 21-23, 1994 Calgary, Alberta

Gibbs, B. F., and Mulligan, C.N. (1997). "Review styrene toxicity: an ecotoxicological assessment" *Ecotoxicology and Environmental safety*" 38,181-194

Chu, W. (2003). "Remediation of Contaminated Soils by Surfactant-Aided Soil Washing" *Practice Periodical of Hazardous, Toxic, and Radio Active Waste Management.* Volume 7, No 1, 19-24

Canadian Environmental Protection Act (1993). *Priority Substances List Assessment Report-* Styrene. Canada Communication Group Publishing, Ottawa Canada

Chemical profile (2004).-Styrene *http://www.the-innovation-group.com/welcome.htm Accessed date 2005.07.26*

Dahr Azma, B. (2002). "Extraction of copper from mining residues by rhamnolipids" *M. A. Sc. Thesis,* Concordia University, Montreal Canada.

DiBacco, D., Hebert, F., Ing, J., Pinchuk, R., and Silverberg, M.(1997). "*Anaerobic degradation of styrene*" McGill University, Department of Chemical Engineering, Project Laboratory report .

Doong, R.A., Lei, W.G., Chen, T.F., Lee, C.Y., Chen, J.H., and Chang, W.H. (1996). "Effect of anionic and non-ionic surfactants on sorption and micellar solubilization of monocyclic aromatic compounds". *Wat Sci and Tech.*34(7), 327-334

Environmental Canada: Summary of Spill Events in Canada: 1984-1995 *http://www.ec.gc.ca/ee-ue/stats/sum_spills/1984_1995/s3_e.asp* Accessed date 2005. 06.01

Environmental Canada NPRI report (2001). *http://www.ec.gc.ca/pdb/npri/2003N_Overview/listFacilities/ON-SITE_RELEASES_e.cfm* Accessed date 2005.05.29

Environmental Quebec: the Land Protection and Rehabilitation Regulation (2003). *http://www2.publicationsduquebec.gouv.qc.ca/dynamicSearch/telechargephp?type=3andfile=/Q_2/Q2R18_1_01_A.htm* Accessed date 2005.06.01

EPA Harris Superfund Site (2004). *http://www.tnrcc.state.tx.us/permitting/remed/superfund/harrisfarleya.html#Process* Accessed date 2005.06.0

EPA Office of Solid Waste and Emergency Response Technology Innovation Office MOTCO Superfund Site (2001). Accessed date 2005.06.01 *http://www.tnrcc.state.tx.us/permitting/remed/superfund/motcoa.html*

EPA Pollution Report NRC# 573731 August 2001. *http://yosemite1.epa.gov/r6/polreps.nsf/0/6b30756e4e79a96786256ab0006be1f9?OpenDocument* Accessed date 2005.06.01

EPA-OPPT chemical fact sheets-Styrene (CAS No. 100-42-5) (1994) *http://www.epa.gov/opptintr/chemfact/styre-sd.txt* Accessed date 2005.05.20

Fu, M. H., and Alexander, M. (1992). "Biodegradation of styrene in samples of natural environments" *Environ. Sci. Technol.* 26.1540-1544

Fu, M. H., and Alexander, M. (1996)." Biodegradation of styrene in waterlogged soils and aquifer solids " *Soil Science.* 161.(12), 846 –851

Huntsman Llc. Styrene Monomer. (2001) *http://www.huntsman.com/base_chemicals/ShowPage.cfm?PageID=1129* Accessed date: 2005.07.15

IPCS INCHEM-Chemical Safety Information from Intergovernmental Organizations. *http://www.inchem.org/documents/pims/chemical/pim509.htm#SubSectionTitle:3.3.3%20%20Description Styrene* Accessed date 2005.05.22

IPCS INCHEM-Chemical Safety Information from Intergovernmental Organizations Environmental Health Criteria 26- styrene 1983. *http://www.inchem.org/documents/ehc/ehc/ehc26.htm* Accessed date 2005.06.01

Kao, C. M., and Lei, S.E. (2000). "Using a peat biobarrier to remediate PCE/TCE contaminated aquifers". *Wat. Res.* 34(3), 835-849

Kao, C.M., Chen, S.C., and Liu, J.K. (2001). "Development of a biobarrier for the remediation of PCE-contaminated aquifer" *Chemosphere* 43,1071-1078

Kao, C.M., Chen, S.C., Wang, J.Y., Chen, Y.L., and Lee, S.Z. (2003). "Remediation of PCE-contaminated aquifer by an in situ two-layer biobarrier: laboratory batch and column studies" *Water Research* 37, 27-38

LaGrega, M.D., Buckingham, P.L., .Evans, J.C., and Environmental Resources Management (2001). "*Hazardous Waste Management*" 2nd Edition McGraw-Hill, Boston, MA.

Liu, J., Olsson, G., and Mattiasson, B. (2003)." Monitoring and Control of an Anaerobic Upflow Fixed-Bed Reactor for High-Loading-Rate Operation and Rejection of Disturbances" *Biotechnology and Bioengineering*, 87(1), 43-53

Liste, H.H., Rivera, A.Q., and Tang, J.X. (2002). "Availability of aged toluene and decane for biodegradation in soil" *Water, Air, and Soil Pollution* 133, 227-234

Massé, D.I., and Massé, L. (2000). "Treatment of Slaughterhouse Wastewater in Anaerobic Sequencing Batch Reactors." *Can. Agric.Eng.* 42(3), 131-137

Myers, D. (1992). "*Surfactant Science and Technology*"2nd edition VCH, New York.

Miller, K.D., Johnson, P.C., Bruce, C.L.(2001). "Full-scale in-situ biobarrier demonstration for containment and treatment of MTBE" *Remediation Journal* 12(1) 25-36

Mulligan, C.N. (2005). "Environmental applications for biosurfactants" *Environmental Pollution* 133 183–198.

Mulligan, C.N., Yong, R.N., and Gibbs, B.F. (2001). "Surfactant-enhanced remediation of contaminated soil: a review" *Engineering Geology* 60 :371-380

Mulligan, C.N. (2002). *Environmental Biotreatment, Technologies for Air, Water, Soil, and Waste* ABS Consulting, Government Institutes, Maryland, USA

National Oceanic and Atmospheric Administration, USA. July 2002. *http://www.coastguard.se/ra/volume2/annex3accidents/styrene.htm* Accessed date 2005.06.01

National Safety Council 2004-styrene chemical backgrounder *http://www.nsc.org/library/chemical/styrene.htm* Accessed date 2005.05.31

Noordman, W.H., Ji, W., Brussear, M. L., and Janssen, D.B. (1998). "Effects of rhamnolipid biosurfactants on removal of phenanthrene from soil" *Environ. Sci. Technol.* 32, 1806-1812

O'Leary, N. D., O'Connor, K. E., and Dobson, A. D.W. (2002). "Biochemistry, genetics and physiology of microbial styrene degradation" *FEMS Microbiology Reviews* 26 403-417.

Pfaffli, P., and Saamanen, A. (1993). "The occupational scene of styrene" *IARC Sci Publ.* (127): 15-26.

Reible, D. D. (1999). "*Fundamentals of Environmental Engineering.*" Lewis Publishers Boca Raton, Fla, USA Shell Chemicals 2005.05 *www.shellchemicals.com/chemicals/pdf/scotford/Report_to_the_Community*.pdf Accessed date 2005.06.01

The Styrene Forum *http://www.styreneforum.org/regulatory_international.html* Accessed date 2005.06.01

ToxFAQs™ for Styrene. ATSDR-Agency for Toxic Substances and Disease Registry *http://www.atsdr.cdc.gov/tfacts53.html* Accessed date 2005.05.23

Tsujii, K. (1998). "*Surface activity, principles, phenomena, and applications*" Academic Press, USA.

Urum, K., Pekdemir, K., and Gopur, M. (2003). "Optimum conditions for washing of crude oil-contaminated soil with biosurfactant solutions" *Trans IchemE*, vol 81 Part B, 203-209

Valdosta State University, Geology 3710, Laboratory 9 (1999). *http://www.valdosta .edu/~grissino/geol3710/lab9.htm*, Accessed date 2005.08.14.

Wang, S.L., and Mulligan, C.N. (2004). " Rhamnolipid foam enhanced remediation of cadmium and nickel contaminated soil". *Water, Air and Soil Pollution* 154:315-330.

Warhurst, A.M., and Fewson, C. A. (1994)." Microbial metabolism and biotransformations of styrene" *J. Appl. Bacteriol.*,77, 597-606

Yerushalmi, L., Manuel, M.F., and Guiot, S.R. (1999). "Biodegradation of gasoline and BTEX in microaerophilic biobarrier" *Biodegradation* 10:341-352.

Yong, R.N., and Mulligan, C.N. (2004). *Natural Attenuation of Contaminants in Soils*. Lewis Publishers Boca Raton, FL, USA.

Yong, R.N., Mohamed, A.M.O., and Warkentin, B.P. (1992). *Principles of Contaminant Transport in Soils.* Elsevier Amsterdam; New York.

In: Hazardous Materials in the Soil and Atmosphere
Editor: Robert C. Hudson, pp. 39-54
ISBN 1-60021-250-6
© 2006 Nova Science Publishers, Inc.

Chapter 2

A STUDY ON THE FEASIBILITY FOR CONTROLLING TRACE METAL LEACHING BEHAVIOR FROM MUNICIPAL SOLID WASTE INCINERATION ASHES

Wu-Jang Huang[1]
National Ping-Tun University of Science and Technology, Department of Environmental Science and Engineering, Ping-Tun, Taiwan

ABSTRACT

In a modern municipal solid waste incineration (MSWI) plant, two air pollution control (APC) residue streams of semi-dry (SD) scrubber residue and fabric filter (FF) ash are generated and usually on-line combined together at a random mixing ratio before for collection. And the collected combined ash is called as MSWI fly ash, in order to distinguish from the coal fly ash. An common phenomena has been recognized in a real MSWI plant is that the toxicity characterization leaching procedure (TCLP) test of a collected MSWI fly ash has a high fluctuation in the leaching concentrations for several easy to leach elements, such as lead and zinc.

In the previous studies [1, 2], the authors have reported the chemical composition analysis of semi-dry (SD) scrubber residue and fabric filter (FF) ash from seven MSWI plants and coleaching behavior of lead from SD/FF mixtures from one of them. In addition, an empirical equation based on the major chemical compositions of original SD residues and original FF ashes has been proposed to describe the coleached concentration of lead from the mix ashes in any ratio [2]. In this study the application feasibility of equation has been extended to three series of synthetic APC residues with varying contents of FF ashes collected from three MSWI plants. And for the first time the fluctuation in TCLP concentration of lead has been suggested that it would be strongly depending on the mixing ratio of FF ash to SD stream mixture, basing on the prediction of our proposed empirical equation. Our data also indicates that the lead (II) concentration of the synthetic APC residues in the TCLP test is lower than the regulatory level, 5.0-ppm, if the mixture contained FF ash less than 50-wt%. If MSW incinerator plant operators blend APC residues with less than 50-wt% of FF ash to consistently

[1] E-mail: wjhuang@mail.npust.edu.tw,TEL. 886-8-7703202 ext. 7076,FAX: 886-8-7740256

produce a non-hazardous and recyclable fly ash under TCLP test, less than 50-wt%FF ash in a collected MSWI fly ash.

1. Introduction

Incineration has become the most important method to reduce the volume of municipal solid waste (MSW) in Taiwan now. Seventeen mass-scale MSWI plants treated over 80% of municipal solid waste incineration (MSWI) in 2001, among them, sixteen plants equipped with a semi-dry (SD) scrubber and a fabric filter (FF) for air pollution control (APC) of the flue gas. In a real plant, two streams of SD residue and FF ash in a MSWI plant are usually collected and mixed together into the stream of APC residues with a random mixing ratio. The collected combined ash is called as MSWI fly ash, in order to distinguish from the coal fly ash. The properties of MSWI fly ash, such as the chemical compositions, the pH of ash, the particle size distribution and the leachability of heavy metals, have been reported well [1~22] over the past twenty years. The total generation rate of combined ash (a mixture of MSWI fly ash and bottom ash) in Taiwan was 2000 tons per day in 2001 and 1800 tons in 2003 [1]. The decrease in generation amount is due to that the people have been asked to recycle reusable matters in MSW before discharged by legislator.

The sprayed amount of lime during the scrubbing step of the flue gas controls the amounts of SD residue and FF ash at the APC line from a MSW incinerator. The amount of sprayed lime changes with operation time and is determined by a computer system relative to the real-time feedback data of the properties of emission gas, such as contents of acid gases and degree of transparence. Therefore, the mix ratio of FF ash in online collected MSWI fly ash indeed is unknown and also difficult to be measured. But one thing must be sure that the major chemical components CaO, SiO_2, and Al_2O_3, in the fly ash changes with time. In addition, an MSWI fly ash is usually enriched with high levels of heavy metals (especially for Pb (II)) and is usually classified as a hazardous material due to it can not pass the toxicity characterization leaching procedure (TCLP) test. A common phenomena seen in most MSWI plants is that an on-line collected MSWI fly ash has a high fluctuation in the concentration of lead in TCLP leachate.

For a same leaching solution in as such the TCLP and the acid neutralization capacity (ANC) tests, the factors affecting the leachability of heavy metals from MSWI fly ash includes the pH of leaching solution, alkalinity, dissolution rate, and sorption process [3]. In our previous studies [1~2], the authors have reported the chemical composition analysis of semi-dry (SD) scrubber residue and fabric filter (FF) ash from seven MSWI plants and coleaching behavior of lead from SD/FF mixtures from one of them. In addition, an empirical equation based on the major chemical compositions of original SD residues and original FF ashes has been proposed to describe the coleached concentration of lead from the mix ashes in any ratio [2]. And in this study the application feasibility of equation will be extended to three series of synthetic APC residues with varying contents of FF ashes collected from three MSWI plants. And for the first time the fluctuation in TCLP concentration of lead has been suggested that it would be strongly depending on the mixing ratio of FF ash to SD stream mixture, basing on the prediction of our proposed empirical equation.

2. Experimental Section

Seven Municipal solid waste (MSW) incineration plants with a total capacity of 900~1350 ton/day were selected to collect their air pollution control (APC) residues. All of them equipped a semi-dry hydrated-lime scrubbing system (de-acid) and after a fabric filter (de-dust) as the cleaning designs for the flue gas. APC residues used in this paper were mainly collected from manhole of fabric filter (referred as *FF ash*) and the bottom of semi-dry scrubber (referred as *SD residue*). The fresh ash was sealed to avoid contacting with CO_2 in the atmosphere and stored for 3 days at a constant temperature of 20~22°C and a moisture of 45~65% before analysis. The fly ash (ie. boiler ash) was not mixed with APC residues.

A Fisher Scientific Accruement pH-meter was used to probe the solution pH value. An Optima 3000XL ion-coupled plasma (ICP) detector made by Perkin-Elmer Co. and a Solar-969 atomic absorption (AA) spectrometer made by Unicam Co. in USA were used to monitor concentrations of species in the final leaching solution. All of the laboratory grade chemicals were purchased from Aldrich Co. and were used directly without purification.

2.1.Toxicity Leaching Extraction Procedure (Tclp) Test

The TCLP test [23] is used as a criterion to determine if a waste is hazardous or non-hazardous. The TCLP test used in this paper followed the USA EPA Method # 1311. Concentrations of heavy metals in the extracts were determined by flame atomic absorption spectrometer following the US EPA Method # 7420. A brief experimental procedure is described below.

Ash samples were graded to pass a 10mm standard sieve. The leaching solution is an acetic acid solution with pH 2.88. The test sample was mixed with acetic acid solution at a liquid/solid ratio of 20.0 c.c. to 1.0 g and agitated at a speed of 30 rpm for 18 hrs. In this paper, the concentration determined using this test was referred as [Pb].

2.2 Preparation Of Binary Mixtures of SD Residue/FF Ash (The Synthetic APC Residues)

Three series of synthetic APC residues were made by mixing the FF ash and SD residue from Plants 1, 2, and 7 at various wt% of dry FF ash. The leaching behaviors of synthetic APC residues also were tested using the TCLP procedure.

2.3 Ph Value Test of Residues

Ten grams of ash sample were mixed with 20-mL $CaCl_2$ standard solution of 0.01M in a 50-mL beaker. The sample was stirred for 30 minutes and stayed for another 30 minutes at room temperature. The pH value of the upper clear liquid solution was measured by pH-meter at 25±2 °C.

2.4 Total Available Concentration Test

Total available concentration test is a standardized method (NVN 7341) to assess the total available concentration of heavy metals, which could leach for 100 years [1, 24]. Ash samples were graded to pass 300um standard sieve. Eight grams of APC residue were extracted by 400 grams of deionized water (pH=7) and 400 grams of nitric acid solution (pH=4). The leachate was adjusted to a liquid/solid ratio of 20 to 1 before measuring.

This method has been slightly modified by using 10% HNO_3 [2] as the leaching solution in TCLP. The ratio between the concentration obtained from the method NVN 7341 and our proposed method [2] was found to be 3.5±0.1 (data not show here). In this paper, the concentration determined using our proposed method is referred to as $[Pb]_o$; while the leachability of lead is referred to as $[Pb]/[Pb]_o$.

2.5 Chemical Composition Analysis by Wet Chemical Analyses

Wet chemical analysis [25] of Al_2O_3, SiO_2, CaO and Fe_2O_3 and CaO/$Ca(OH)_2$ was determined using ASTM Method C114-94. The analytical procedures used the Standard Testing Methods of Republic of China [6]. The levels of major components of Al_2O_3, SiO_2, CaO and Fe_2O_3, and miner components of free CaO and $Ca(OH)_2$ were determined basing on differential wet-chemistry In the major component part, a 0.5-g dried sample was extracted in concentrated HCl and HNO_3 in an addition of 0.5g NH_4Cl several times. The insoluble fraction was dried and weighted; it is attributed to the sum of SiO_2, $Al(OH)_3$ and Fe_2O_3. The residual filtrate was reheated, and NH_4NO_3 added, and the precipitate was collected after filtration. The filtered solution is used for determined (CaO+ $Ca(OH)_2$)% and the weight of precipitated NH_4OH is used to calculate the $Al(OH)_3$% later. Titration of $KMnO_4$ solution with the final filtrate of above step is an optional method to the minor component in this method. The percentage of Fe_2O_3 was determined by titration of $K_2Cr_2O_7$ using a fresh sample dissolved in concentrated HCl and oxidized by $SnCl_2$(aq). The difference of weight of NH_4OH and Fe_2O_3 is the weight of $Al(OH)_3$ and all compositions of residues are available.

2.6 Instrumental Analysis

A Fisher Scientific Accruement pH-meter was used to probe the pH value of the solution. A Solar-969 atomic absorption (AA) spectrometer made by Unicam Co. USA, was used to monitor concentrations of heavy metals. The reagent-grade chemicals were obtained from commercial products and were used directly without further purification.

3. Results and Discussions

The APC residue of any mass-scale MSWI plant usually was on-line produced with a random mixing ratio of MSW SD residues and MSW FF ashes. In the present study, the authors individually characterized the chemical compositions and the leachability of lead of

MSW SD residues and MSW FF ashes. Three series of synthetic APC residues were prepared through adding various weight percentages of MSW FF ashes into MSW SD residues. Varying the ratios of MSW FF ashes in synthetic APC residues changes their chemical composition.

3.1 Chemical Composition Analyses of Semi-Dry (SD) Scrubber Residue and Bag-House Ash

Several research groups have reported [1-3, 10,11,17-20] the chemical composition of APC residues from MSW incinerators equipped with a semi-dry scrubber and a bag-house (fabric filter, FF) as the gas-cleaning system. The International Ash Working Group has reported studies on SD residues and FF ashes from MSWI [3]. Table 1 and Table 2 summarized chemical composition reported in the literatures for MSW SD and FF ashes, respectively. In semidry scrubber/FF systems, the SD residues have higher contents of calcium, silicate and aluminate, while FF ashes have higher content of calcium. This paper compares data from this study with literature results of residues from semidry scrubbers and bag houses, not with that of wet scrubbers and electrostatic precipitators.

The authors have analyzed the main chemical composition of the SD residues and FF ashes used in this study [1] and summarized them in Table 3, for which the reported content of CaO might include $CaCO_3$ and $CaSO_4$. Based on Table 3, SD residues possess a higher content of SiO2 than FF ashes and FF ashes possess a higher content of CaO. Moreover, SD residues have a narrower distribution for the ratio of CaO/SiO2, being close to 1.0 ± 0.3, and a broader distribution on the ratio of CaO/Al2O3, ranging from 0.6 to 3.20. In contrast, FF ashes have a sharper distribution on the ratio of CaO/Al2O3 0.3 ± 0.2, and a broader distribution on the ratio of CaO/SiO_2, with a rang of 2.4 to 13.2. These ratios are consistent with the results of the International Ash Working Group [3]. In this paper, the content of Al_2O_3 in APC ashes are higher than that reported for the USA or Europe [3] and might result from the un-recycled aluminum matter in the raw MSW for Taiwan.

Table 1. Chemical Composition of MSWI Scrubber Residue from Literatures

Sources	Type of APC	Si or SiO_2 (wt %)	Al or Al_2O_3 (wt %)	Ca or CaO (wt %)	Ca/Si	Ca/Al
Ref. [9]	ESP+Dry scrubber	19	6.5	8.6	0.45	1.32
Ref. [10]	Semi dry scrubber +FF	15.0	7.07	27.7	1.85	3.92
Ref. [17]	Semi dry scrubber +FF	0.1	4.89	47.3	473	9.67

Table 2. Chemical Composition of MSWI Precipitator Ash from Literatures

Sources	Type of APC	Si or SiO_2 (wt %)	Al or Al_2O_3 (wt %)	Ca or CaO (wt %)	Ca/Si	Ca/Al
Ref. [10]	Semi dry scrubber +FF	18.76	12.74	24.3	1.30	1.91
Ref. [16]	ESP+Dry scrubber	4.8	2.6	5.8	1.21	2.23
Ref. [17]	Semi dry scrubber +FF	0.2	9.1	37.8	189	4.15
Ref. [18]	ESP+Wet scrubber	30	13	29.7	0.99	2.28
Ref. [19]	ESP+Dry scrubber	10	4.3	25.2	2.52	5.86

Table 3. The Chemical Compositions of APC Residues in the West of Taiwan [1]

Pants	SiO_2 (wt%)		Al_2O_3 (wt%)		CaO (wt%)	
	FF ash	SD residue	FF ash	SD residue	FF ash	SD residue
1	9.26	32.20	26.33	12.0	55.21	38.45
2	9.12	20.60	16.15	39.31	50.75	26.71
3	17.57	19.60	32.30	46.31	42.36	32.20
4	3.12	36.48	42.82	32.63	41.39	26.25
5	3.76	21.08	37.48	44.71	45.65	26.14
6	6.80	24.28	16.40	33.59	55.20	38.28
7	3.91	36.68	37.93	24.61	42.60	28.16

CaO/SiO2		CaO/Al2O3	
FF ash	SD residue	FF ash	SD residue
5.96	1.19	0.35	3.20
5.56	1.30	0.56	0.68
2.41	1.64	0.54	0.69
13.27	0.72	0.19	0.80
12.14	1.24	0.10	0.58
8.12	1.58	0.41	1.14
10.90	0.77	0.10	1.14

* For detail experience please see Ref. [1] .

3.2 Leaching Toxicity of SD Residues and FF Ashes

In the present study, the authors have focused on the leachability of lead (II) due to its high leachability from the APC residue [3]. In our previous study [1], the authors demonstrated that SD residue and FF ash are cement-like materials. Besides this, Pb (II) leachability increases with the CaO/SiO_2 ratio in FF ash and with the CaO/Al_2O_3 ratio in SD residue.

Table 4 shows results of ash pH, concentrations of lead ([Pb]) and chloride (Cl^-) and the final pH in the final TCLP leaching solution of SD residues and FF ashes. In present paper, $[Pb]/[Pb]_o$ defines the leachability of lead(II). The $[Pb]_o$ represents the total available concentration of lead, and the [Pb] represents the result from TCLP test, both collected in the same solution to mass ratio and test standard for TCLP (please see the "*Experiments*" section). This definition of $[Pb]/[Pb]_o$ allows comparison with varying lead contents and TCLP performance of the relative leachability between residues from various sources. A waste could be defined as a hazardous waste, if it did not pass the TCLP test. Each of the tested FF ashes is a hazardous waste from the results in Table IV due to high lead concentration. In contrast, SD residues show a lower concentration of lead. In addition to lead, concentrations of Cd(II), Cr(III), Cr(VI), Zn(II), Cu(II) were found to be far below their regulation levels with respect to the law of Taiwan's Agency of Environmental Protection (data not show here).

Table 4. Properties of Final TCLP Leaching Solutions of FF Ashes and SD Residues

Plants	FF ashes			SD residues		
	$[Pb]_o$ (ppm)	[Pb] (ppm)	$[Pb]/[Pb]_o$	$[Pb]_o$ (ppm)	[Pb] (ppm)	$[Pb]/[Pb]_o$
1	115.2	12.76	0.110	27.8	2.587	0.0930
2	72.86	11.2	0.154	20.44	1.24	0.0606
3	126.7	0.11	8.69E-4	31.1	0.261	0.0084
4	263.5	73.30	0.278	133.9	0.696	0.0052
5	147.6	38.04	0.258	22.4	0.065	0.0029
6	79.1	9.72	0.123	27.0	0.63	0.0233
7	172.8	38.61	0.223	17.0	0.46	0.0271

Plants	Ash pH		pH of Final TCLP Leachate		$[Cl^-]$	
	FF ash	SD residue	FF ash	SD residue	FF ash	SD residue
1	12.11	11.86	12.38	12.23	6000	1130
2	12.15	11.40	12.14	7.85	6800	1710
3	12.08	11.47	10.53	7.55	6430	4230
4	12.14	11.68	12.04	11.61	13140	4970
5	12.11	12.01	12.35	12.10	8980	1480
6	12.09	11.60	12.35	12.10	6550	3440
7	12.10	11.89	12.29	11.38	10260	1600

Fig. 1 and Fig. 2 show the relationship of the lead leachability with the ratios of CaO/SiO_2 of FF ashes and CaO/Al_2O_3 of SD residues, respectively. Linear regression of the data in Figures 1 and 2 results in Equation (1) and (2), respectively.

$$([Pb])_{FFash} = ([Pb]_o)_{FFash} * \{-0.0692 + 0.0263 \times (\frac{CaO\%}{SiO_2\%})_{FFash}\} \quad (1)$$

$$([Pb])_{SDresidue} = ([Pb]_o)_{SDresidue} * \{-0.0162 + 0.0342 \times (\frac{CaO\%}{SiO_2\%})_{SDresidue}\} \quad (2)$$

The intercept value on the x-axis and y-axis is near 2.3 and 0.5 for Fig. 1 and for Fig. 2, respectively. This implies that lead leachability increases as above CaO/SiO_2 ratio of 2.3 and CaO/Al_2O_3 ratio of 0.5. As shown in Table 3, the FF ash samples had ratios of $CaO/Al_2O_3 < 0.5$, always the lead leachate concentrations of FF ashes are almost insensitive with regard to the ratio of CaO/Al_2O_3. Similarly, all of the SD residue samples have ratios of $CaO/SiO_2 < 2.0$, implying that the lead leachate concentrations of SD residues are almost insensitive with regard to the ratio of CaO/SiO_2.

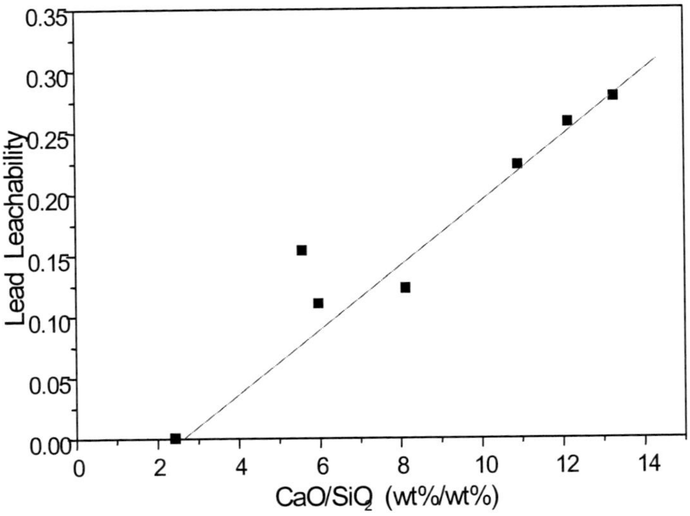

Figure 1. Plot of ratio of CaO/SiO_2 versus leachability of lead(II) of FF ashes. [1]

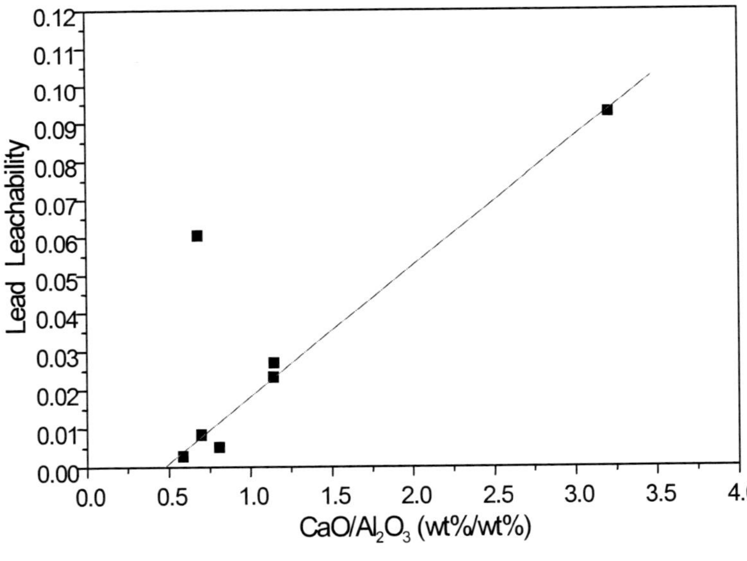

Figure 2. Plot of ratio of CaO/Al_2O_3 versus leachability of lead(II) of SD residue samples. [1]

3.3 The Leaching Behaviors of the Synthetic APC Residues (Binary Mixtures of SD Residues and FF Ashes)

The authors were interested in the leaching behavior of binary mixtures of FF ashes and SD residues; to address the high fluctuation on the lead leachate concentrations of APC residues in the TCLP test. Three series of synthetic APC residues were prepared by mixing each collected FF ash and SD residue from Plant 1, 2, and 7. Fig. 3~5 and Table 5 show concentrations of lead in the TCLP leaching solution from these synthetic APC residues.

Each system shows *a negative deviation on* [Pb] concentration upon adding SD into the FF ashes. According to the leaching mechanisms of lead from APC residues of MSWI plants [2, 5], this phenomenon results from a systematical change of the pH values of the synthetic APC residues or there are specific interactions between lead ions and particulate in the SD residue.

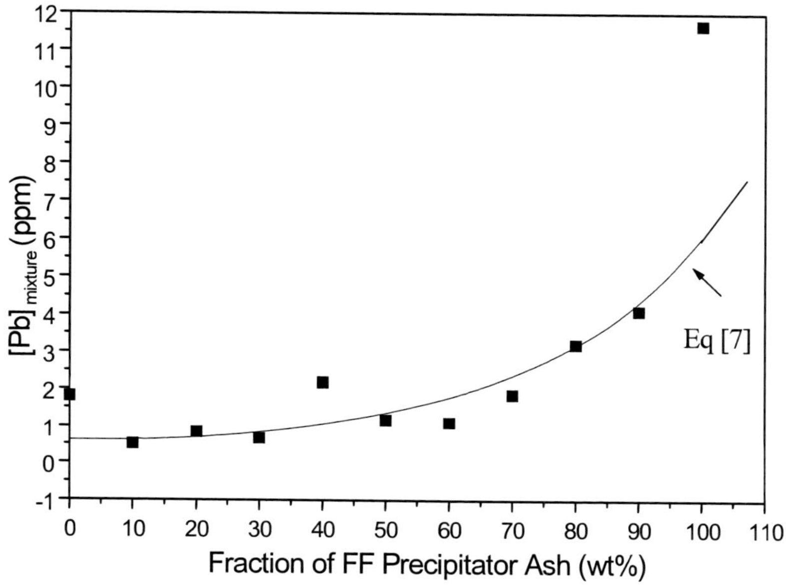

Figure 3. Plots of TCLP leaching data of lead of the synthetic ashes from Plant 1 and their predicted data from Eq. (7).

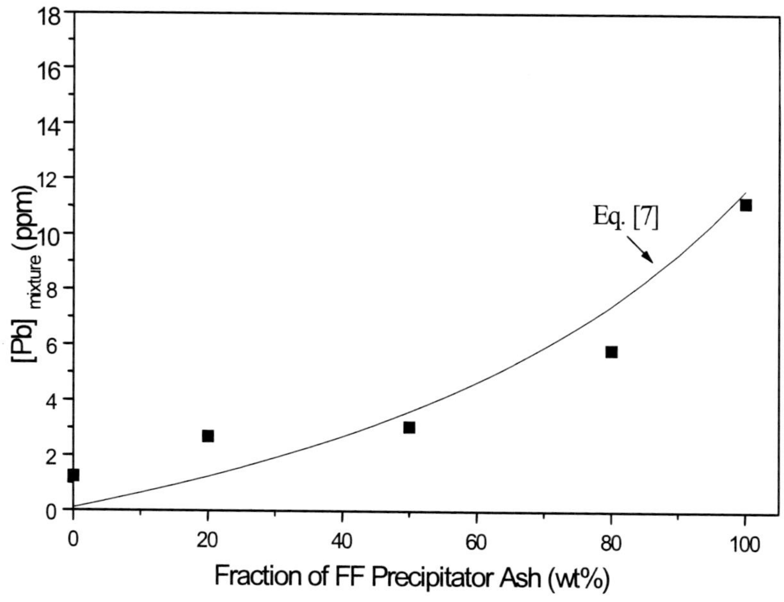

Figure 4. Plots of TCLP leaching data of lead of the synthetic ashes from Plant 2 and their predicted data from Eq. (7).

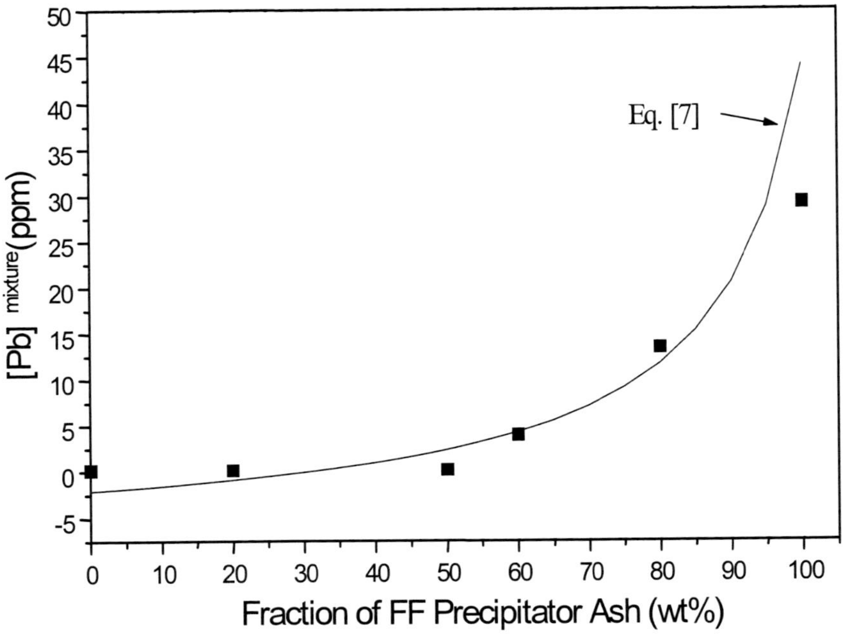

Figure 5. Plots of TCLP leaching data of lead of the synthetic ashes from Plant 7 and their predicted data from Eq. (7).

Table 5. TCLP Data of the Synthetic APC Residues

Wt% of FF ash / [Pb] (ppm)	Plant 1	Plant 2	Plant 7 [2]
100	11.63	11.2	29.175
90	4.10	-	-
80	3.22	5.85	13.4
70	1.85	-	-
60	1.08	-	3.95
50	1.15	3.08	0.175
40	2.18	-	-
30	0.65	-	-
20	0.83	3.2	0.079
10	0.50	-	-
0	1.8	1.24	0.211

Combining Eq. (1) and Eq. (2) through simple additive contribution of lead leachability for each component of the binary mix. This equation also considered their weight fractions, f FF ash and f SD residue.

$$[\mathrm{Pb}]_{\mathrm{mixture}} = [\mathrm{Pb}]_{\mathrm{FF\ ash}} + [\mathrm{Pb}]_{\mathrm{SD\ residue}} \quad (3)$$

$$[\mathbf{Pb}]_{\mathrm{mixture}} = f_{\mathrm{FF\ ash}} * \{-0.0692 + 0.0263 \times (\frac{CaO\%}{SiO_2\%})_{FFash}\}$$

$$+ f_{SD\ residue}{}^{*}\{-0.0162 + 0.0342 \times (\frac{CaO\%}{Al_2O_3\%})_{SDresidue}\} \quad (4)$$

Eq. (4) does not fit the results of Table 5 well because the chemical composition of the combined mixture (synthetic APC residues) was not considered within Eq. (4). Modifying CaO/SiO_2 ratio of the FF and CaO/Al_2O_3 ratio of the SD to rhe respective ratios for the mixture resulted in Eq. (5). Ratios of C CaO/SiO_2 and CaO/Al_2O_3 of a combined mixture can be calculated using Eq. (6).

$$[Pb]_{mixture} = ([Pb]_o)_{FFash} * \{-0.0692 + 0.0263 \times (\frac{CaO\%}{SiO_2\%})_{mixture}\}$$

$$+([Pb]_o)_{SDresidue} * \{-0.0162 + 0.0342 \times (\frac{CaO\%}{Al_2O_3\%})_{mixture}\} \quad (5)$$

$$(\frac{CaO\%}{SiO_2\%})_{mixture} = \frac{f_{FF\ ash} \times (CaO\%)_{FF\ ash} + f_{SDresidue} \times (CaO\%)_{SDresidue}}{f_{FF\ ash} \times (SiO_2\%)_{FF\ ash} + f_{SDresidue} \times (SiO_2\%)_{SDresidue}}$$

$$(\frac{CaO\%}{Al_2O_3\%})_{mixture} = \frac{f_{FF\ ash} \times (CaO\%)_{FA} + f_{SDresidue} \times (CaO\%)_{SDresidue}}{f_{FF\ ash} \times (Al_2O_3\%)_{FA} + f_{SDresiduee} \times (Al_2O_3\%)_{SDresidue}} \quad (6)$$

$$f_{FF\ ash} + f_{SD\ residue} = 1$$

Authors added another four parameters of S_1, S_2, C_1, and C_2 in Eq. (7). Among them, S_1 and S_2 represented the degrees of sorption of lead of a FF ash and a SD residue, and C_1 and C_2 represented the effective leaching fraction of lead(II) of a FF ash and a SD residue, respectively.

$$[Pb]_{mixture} = S_1{}^{*}\{([Pb]_o)_{FFash} * \{-0.0692 + C_1 * 0.0263 \times (\frac{CaO\%}{SiO_2\%})_{mixture}\}\}$$

$$+ S_2{}^{*}\{([Pb]_o)_{SDresidue} * (-0.0162 + C_2 * 0.0342 \times (\frac{CaO\%}{Al_2O_3\%})_{mixture}\} \quad (7)$$

Eq.(7) fits the data Fig. 3-5 well by using the fitting parameters listed in Table 6. All S_2 and C_2 were found to be near 1.0, each S_1 was smaller than 1.0, and each C_1 was 1 or 2.

3.4 The Leaching Mechanism of the Synthetic Ashes

In Tables 7 to 9, the authors summarized the final pH and contents of lead (Pb), silicon (Si), calcium (Ca), aluminum (Al) sodium (Na), and potassium (K) in the final leaching solution for Plants 1, 2 and 7. The Si and Al ions leached from the three synthetic ashes were very low, indicating that silicate and aluminate salts have less solubility in the acetic acid-based leaching solution.

Table 6 Fitted Parameters of the Synthetic APC Residues

Plants	S_1	S_2	C_1	C_2
1	0.45	1.0	1.0	1.0
2	0.60	1.0	2.0	1.0
7	0.50	1.0	2.0	1.0

Additionally, the leaching behaviors of lead from the synthetic Plant 1 and Plant 7 ashes are neither a pH-dependent nor a Ca-dependent. As mentioned above, the final pH of the leaching solution might be dependent on the content of CaO and Ca $(OH)_2$ in the residues and being the important controlling factor on lead leachability. The results show that increasing the FF content in a synthetic ash actually increases the final pH of leaching solution linearly, but not proportionally increase the lead leachability for FF content less than 50-wt%. Besides the direct dissolution process (pH controlled only), the factors such as sorption, adsorption or diffusion shall be considered together for lead leaching.

In the report of Cheng and Bishop [27], silicon was the dominat species in the TCLP leachants of the synthetic heavy metal containing concrete by cement solidification, and they concluded that the sorption mechanism of metals in solution should be the controlling factor. Baker and Bishop [28] reported the shrinking un-reacted core model to describe the retarded leaching behavior of lead. In that model, the inward diffusion of acid species into the alkaline depleted leached shell controls the leaching rate of lead mainly, but with the adsorption of lead onto the leached shell as an important secondary mechanism, in the kinetic controlled release mechanism of lead from the matrix.

Barna et al. [29] used the bulk diffusion model to describe the leaching curves of Na^+ and K^+ from solidified soil. They concluded that the coupled dissolution-diffusion model was more suitable for Ca^{2+} and Pb^{2+} and the leaching of lead was mainly the dissolution process due to its high surface sensitivity. Baur et. al [30] also conducted the leaching experiments on cement solidified air pollution control residues, and confirmed that the leaching mechanism of lead is a dissolution process not a diffusion process. Our previous report also pointed out that the surface adsorption process of lead is important in determining the diffusivity of lead ions in a solidified concrete [31].

At present, although the leaching mechanism of lead from the MSWI residues can not fully be explained considering only a few factors, but this work clearly demonstrated the weak effects of extractant pH and calcium content on the lead leachability of a series of synthetic ash mixtures. Therefore, the dissolution-diffusion-adsorption processes may control the final leaching concentration of lead from the ash, and its major chemical compositions affect its metal adsorption capacity.

In the shrinking un-reacted core model proposed by Baker and Bishop [28], the total amount of lead in the matrix depends on the content of acid available. In this paper, the total metal content is considered significant to the leaching mechanism, but the available acid is also considered significantly, thus, the total amount of lead is not reported. (**This statement is only true upon that acid is in a limited supply condition, such as the waste leaching by acid rain or by the acidic solutions in a TCLP testing jar**).

Table 7. Properties of Final TCLP Leaching Solution of Synthetic Ashes from Plant 1

% of FF	pH	Pb (ppm)	Ca (ppm)	Si (ppm)	Al (ppm)	Na (ppm)	K (ppm)
100	12.14	11.63	5080	ND	ND	1760	1180
90	12.04	4.10	5000	ND	ND	1710	1180
80	12.02	3.22	4780	ND	ND	1690	1140
70	11.88	1.85	4400	ND	ND	1890	1130
60	11.14	1.08	4250	ND	ND	1680	1100
50	10.34	1.15	3970	ND	ND	1630	944
40	9.92	2.18	3900	ND	ND	1670	1020
30	9.33	0.65	3590	ND	ND	1380	1000
20	8.54	0.83	2400	0.72	ND	2080	1150
10	8.26	0.50	2270	66.8	ND	1690	996
0	7.854	1.8	1870	80.4	5.98	1460	829

* ND: Lower than detection limit.

Table 8. Properties of Final TCLP Leaching Solution of Synthetic Ashes from Plant 2

% of FF	pH	Pb (ppm)	Ca (ppm)	Si (ppm)	Al (ppm)	Na (ppm)	K (ppm)
100	12.38	11.2	3070	ND	ND	1870	1265
80	12.34	5.85	2780	ND	ND	1840	1140
50	12.31	3.08	2540	ND	ND	1520	1050
20	12.30	3.2	2000	4.91	ND	1460	912
0	12.23	1.24	1560	82.4	0.95	1610	863

* ND: Lower than detection limit

Table 9. Properties of Final TCLP Leaching Solution of Synthetic Ashes from Plant 7

% of FF	pH	Pb (ppm)	Ca (ppm)	Si (ppm)	Al (ppm)	Na (ppm)	K (ppm)
100	12.29	29.18	3030	ND	0.01	2300	1140
80	12.15	13.40	2880	ND	0.01	1930	975
60	12.04	3.95	2700	ND	0.01	1510	809
50	11.98	0.18	2610	ND	0.02	1330	715
20	11.71	0.08	2190	1.04	0.11	779	429
0	11.38	0.21	1920	18.2	0.09	421	226

*ND: Lower than detection limit.

These results provide an explanation for the high fluctuation of the TCLP test with sampling time in on-line generated APC residues with the SD residues randomly mixed with the FF ashes. The equation could be useful to MSW incinerator plant operators for blending APC residues consistently to produce a fly ash deemed non-hazardous under Taiwan's environmental law. Most of the on-line collected APC residues were stabilized by cement based solidification treatment before being disposed of in landfills, however, the mixing ratio of APC residues to cement is operated at a constant feed rate, without any consideration for mixing ratio of FF to SD. If the on-line generated APC residue is non-hazardous, it could be recycle used.

4. Conclusions

This study characterized the chemical composition and the leaching behaviors of lead (II) of two APC residues from MSWI plants, SD residues and FF ashes, and their binary mixtures. Using the TCLP leaching test, lead (II) starts to leach significantly at a CaO/SiO_2 ratio > 2.3 for SD residues and a CaO/Al_2O_3 ratio > 0.5 for FF ashes. TCLP leachate concentrations of lead of three series of synthetic APC residues increased significantly for ashes with 60wt% FF ash. This result provides an explanation for the high fluctuation in lead leached from on-line generated APC residues.

Acknowledgement

The author would like to thank the National Science Council of Republic of China (TAIWAN) provided finical support of this research under a contract number NSC 92-2113-M-020-001.

References

[1] W.-J.Huang and S.-C. Chu, A study on the cement-like property of municipal incineration ashes, *Cem. Concre. Res.* 33 (2003) 1795-1799.

[2] W.-J. Huang and S.-C. Chu, Coleaching behavior of lead of incineration ash waste mixtures, *Cem. Concre. Res.* 35 (2005) 1038-1041.

[3] "Municipal solid waste incinerator residues-volume I", Ed by The international Ash Working Group, Elsevier, New York, *Characterization of air pollution control residues*, Chapter 11 (1997) 441-482.

[4] S. V. Mattigod, D. Rai, L. E. Eary and C. C. J. Ainsworth, Geochemical factors controlling the mobilization of inorganic constitutes from fossil fuel combustion residues: I. *Review of the major elements, Environ. Qual.* 19 (1990) 188-201.

[5] Kida, Y. Noma, and T. Imada, Chemical speciation and leaching properties of elements in municipal incinerator ashes, *Wast. Manage.* 16 (1996) 527-536.

[6] F. P. Glasser, Fundamental aspects of cement solidification and stabilization, *J. Hazard. Mater.* 52 (1997) 151-170.

[7] H. Belevi and H. Moench, Factors Determining the Element Behavior in Municipal Solid Waste Incinerators. 1. Field studies, *Environ. Sci. Technol.* 34 (2000) 2501-2506.
[8] T. T. Eighmy, J. D. Jr. Eusden, J. E. Krzanowski, D. S. Domingo, D. Stampfli, J. R. Martin and P. M. Erickson, Comprehensive approach toward understanding element speciation and leaching behavior in municipal solid waste incineration electrostatic precipitator ash, *Environ. Sci. Technol.* 29 (1995) 629-646.
[9] S. H. Lee, E. Sakai, M. Daimon and W. K. Bang, Characterization of Fly Ash directly Collected from Electrostatic Precipitator, *Cem. Concre. Res.* 29 (1999) 1791-1797.
[10] E. C. Mcdonald-Buller, I. W. Duedall, E. H. Kalajian, and S.-C. Shien, Measurement and Stabilization of Waste-to-Energy Ash Concrete Expansion in Seawater, *Environ. Sci. Technol.* 34 (2000) 3211-3216.
[11] N. Alba, E. Vazquez, S. Gasso and J. M. Baldasano, Stabilization/solidification of MSW incineration residues from facilities with different air pollution control systems. Durability of matrices versus carbonation, *Wast. Manag.* 21 (2001) 313-323.
[12] T. T.Eighmy, B. S.Crannell, L. G. Butler, F. K. Cartledge, E. F. Emery, D. O. James, E. Krzanowski, J. D. Jr. Eusden, E. L. Shaw and C. A. Francis, Heavy Metal Stabilization in Municipal Solid Waste Combustion Dry Scrubber Residue using Soluble Phosphate, *Environ. Sci. Technol.* 31 (1997) 3330.-3338
[13] Y. Wang, D. Ren, and F. Zhao, Comparative Leaching Experiments for Trace Elements in Raw Coal, Laboratory Ash, Fly Ash and Bottom Ash, *Intern. J. Coal Geo.* 40 (1999) 103-108.
[14] Hauser, U. Eggenberger, and T. Peters, Original and characterization of fly ash from cellulose industries containing high proportions of free lime and anhydrite, *Cem. Concre. Res.* 29 (1999) 1569-1573.
[15] P. Arjunan, M. R. Silsbee, and D. M. Roy, Sulfoaluminate-belite cement from low calcium fly ash and sulfur-rich and other industrial by-products, *Cem. Concre. Res.* 29 (1999) 1305-1311.
[16] J. R. Connor, *Chemical Fixation Solidification of Hazardous Wastes*, Van Nostrand-Reinhold, New York (1990).
[17] T. T. Eighmy, I. D. Jr. Eusden, J. E. Krzanowski, D. S. Domingo, D. Stampfli, J. R. Martin and P. M. Erickson, Comprehensive approach toward understanding element specification and leaching behavior in municipal solid waste incineration electrostatic precipitator ash, *Environ. Sci. Techn.* 29 (1995) 629-646.
[18] K. Lundtorp, D. L. Jensen, M. A. Sorensen, H. Mosbaek and T. H. Christensen, On site treatment and landfilling of MSWI air pollution control residues, *J. Hazard. Mater.* B97 (2003) 59-70.
[19] F. Medici, L. Piga and G. Rinald, Behavior of polyamniophenolic additives in the granulation of lime and fly-ash, *Wast. Manag.* 20 (2000) 491-498.
[20] Polettini, R. Poni, P. Sirini and F. J. Testa, Properties of Portland cement-stabilized MSWI fly ashes, *Hazard. Mater.* B88 (2001) 123-138.
[21] J.G. S. Van Jaarsveld and J.S.J. van Deventer, Effect of the alkali metal activator on the properties of fly ash-based geopolymers, *Ind. Eng. Chem. Res.* 38 (1999) 3932-3941.
[22] J. Yan and I. Neretnieks, Is the glass phase dissolution rate always a limiting factor in the leaching processes of combustion residues, *The Sci. Total. Environ.* 172 (1995) 95-118.

[23] "Analysis of Hazardous Wastes", US EPA Methods # 1131 and 7420.

[24] "Municipal solid waste incinerator residues-volume II", Ed by The international Ash Working Group, Elsevier, New York, 1997, Leaching tests, Chapter 14 (1997) 579-606.

[25] "Chemical Compositional Analysis of Portland Materials", Method of ASTM C114-94.

[26] "Municipal solid waste incinerator residues-volume II", Ed by *The international Ash Working Group*, Elsevier, New York (1997) 938-941.

[27] K. Y. Cheng, and P. L. Bishop, Sorption, important in stabilized/solidified waste forms, Hazard. *Wast. and Hazard. Mater.* 9 (1992) 289-296.

[28] P. G. Baker, and P. L. Bishop, Prediction of metal leaching rates from solidified/stabilized containing wastes using the shrinking unreacted core leaching procedure, *J. Hazard. Mater.* 52 (1997) 311-333.

[29] F. Sanchez, R. Barna, A. Garrbrants, D. S. Kosson and P. Moszkowicz, Environmental assessment of a cement-based solidified soil contaminated with lead, *Chem. Eng. Sci.* 55 (2000) 113-128.

[30] I.Baur, C. Ludwig and J. C. Annette, The leaching behavior of cement stabilized air pollution control residues: A comparison of field and laboratory investigation, *Environ. Sci. and Techn.*35 (2001) 2817-2822.

[31] W.-J. Huang, Y.-C. Huang and C.-C. Lee, Pb(II) diffusivity in cement solidified MSWI fabric filter precipitator ash with a column leaching test measurement, *J. Chine. Instit. Environ. Eng.* 13 (2003) 175-182.

In: Hazardous Materials in the Soil and Atmosphere
Editor: Robert C. Hudson, pp. 55-88
ISBN 1-60021-250-6
© 2006 Nova Science Publishers, Inc.

Chapter 3

ALTERNATIVE HALOMETHYLATION PROCESSES AND THE CORRESPONDING PREPARATION ROUTES FOR ANION EXCHANGE MEMBRANES BY AVOIDING THE USE OF HAZARDOUS MATERIALS CHLOROMETHYL METHYL ETHER (CME) AND BIS-CHLOROMETHYLETHER (BCME)

Tongwen Xu[1]
Lab. of Functional Membrane, School of Chemistry and Material Science,
University of Science and Technology of China, Hefei, P.R. China

ABSTRACT

Ion exchange membranes have been widely used in various industrial fields, such as electrodialysis to concentrate or deionize aqueous electrolyte solutions, diffusion dialysis to recover acid or alkali from waste acid or alkali solutions, anion exchange layer constructed into a bipolar membrane for cleaning productions and separations, polymer electrolyte for fuel cell, etc. To satisfy with these requirements, strong acid type cation exchange membranes with sulfuric group and strong base type anion exchange membrane with quaternary ammonium group were commonly prepared from styrene-divinylbenzene copolymers. In such a route, preparation of a cation exchange membrane seems to be easier by simple sulphonation with concentrated sulfuric acid or chloride sulphonic acid; while preparation of an anion exchange membrane had to experience more complicated procedures: chloromethylation and quaternary amination. Particularly, in the chloromethylation, the used chloromethyl methyl ether or bis-chloromethylether is a potent carcinogen and its use has been restricted since the 1970s. Till now, some efforts have been made to cancel the use of such hazardous material. Therefore, in this chapter, these routes for anion exchange membrane preparations in the absence of chloromethyl methyl ether or bis-chloromethylether will be reviewed and commented.

[1] E-mail: twxu@ustc.edu.cn; tel:+86-551-3601587; fax:+86-551-3601592

Keywords: anion exchange membrane; chlomethylation;halomethylation; harzadous meterials

1. INTRODUCTION

Ion exchange membranes have found many applications ever since they were invented in the last century. For examples, they can be used in electrodialysis to concentrate or dilute in aqueous or non-aqueous electrolyte systems, and in diffusion dialysis to recover acid or alkali from waste acid or alkali solutions [1-5]. They can also separate acidic gas via carrier transport [6-8], and act as solid electrolytes in fuel cells [9,10] or as sensing materials in sensors [11]. To satisfy with the demands, strong acid type cation exchange membranes with sulfuric group and strong base type anion exchange membrane with quaternary ammonium group were commonly prepared from copolymers or polymers [12-13]. Compared with cation exchange membranes, which is simply prepared by benzyl sulphonation with concentrated sulfuric acid or chloride sulphonic acid, preparation of an anion exchange membrane had to experience more complicated procedures: chloromethylation and quaternary amination. Particularly, in the chloromethylation, the commonly used chloromethyl methyl ether (CME) and bis-chloromethylether (BCME), which provide excellent conversions and high yields, are now considered to be carcinogens and their use has been restricted since the 1970s [14-16]. Therefore, anion exchange membrane preparation route free from CME or BCME are strongly recommended. Till now, many efforts have been made to seek alternative methods ever since, such as: amidoalkylation of crosslinking polystyrene[17], copolymerization of (chloromethyl)styrene [18], halogenation of polymers[19-20], chloroacetylation of polymers [21-22], acetylation of crosslinked polystyrene[19], grafting vinyl benzyl trimethyl ammonium chloride onto inert polymers[23], etc.. Therefore, in this chapter, the detail preparation route as well as the fundamental results will be described in the following sections.

2. ANION EXCHANGE MEMBRANES PREPARATION BY BROMINATION PROCESS

It is worthy of mention, recently, our lab has developed a new route, in which the anion exchange membranes were directly made from engineering polymer poly (2,6-dimethyl-1,4-phenylene oxide) (PPO) by bromination and quaternary amination instead of chloromethylation with chloromethyl methyl ether (As shown in Fig. 1b for comparison with conventional route Fig. 1a). The advantages of such route are that the properties of the membranes can not only be quantitatively controlled by bromination process (such as the substitution position or content) but also be controlled by the subsequent amination-crosslinking processes [24-25]. As examples, Table 1 gives the properties change of membranes with benzyl and aryl substitutions. Obviously, for the first membrane series, IEC are enhanced by the benzyl substitution Nevertheless, fixed group concentration, which are expressed as the ratio of IEC to water content, decreased at first and remain approximately unchanged due to the increase in water content.

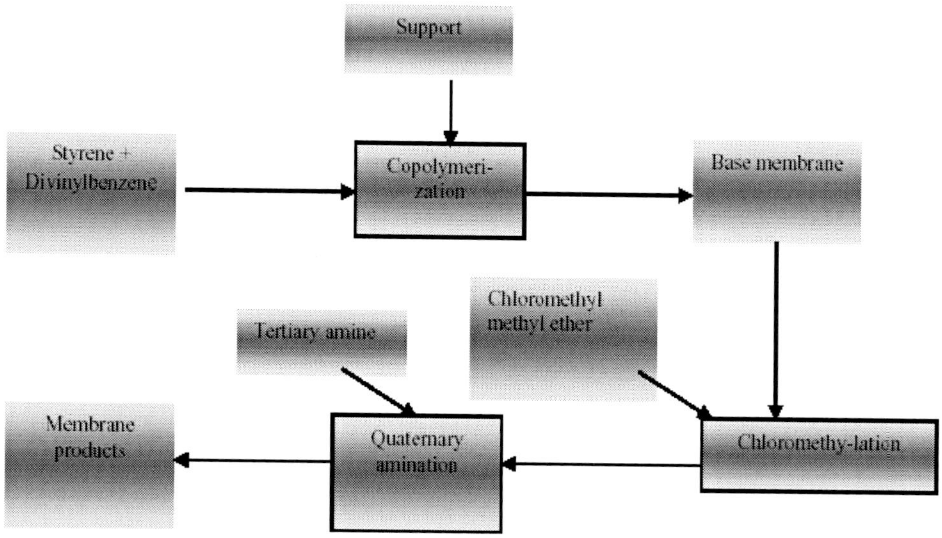

(a) Route for a traditional hydrocarbon type anion exchange membrane

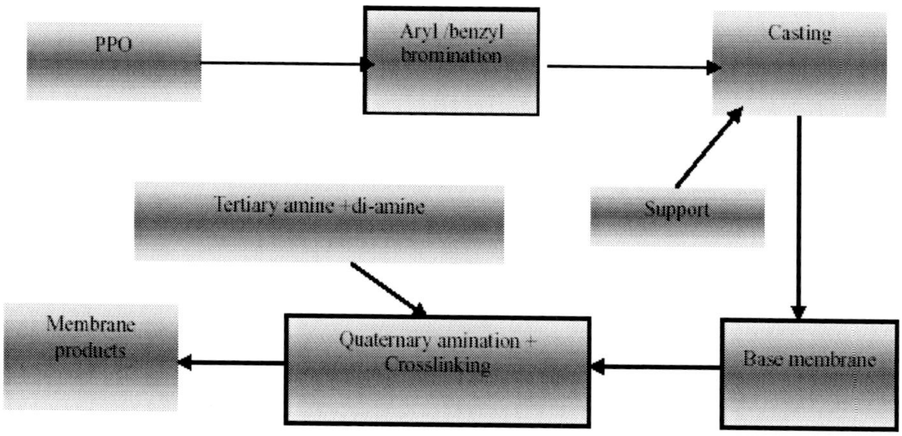

(b) Route for a new anion exchange membrane prepared from PPO by bromination and quaternary amination processes

Fig. 1. A comparison in manufacture routes between conventional hydrocarbon type anion exchange membranes and the anion exchange membranes prepared from polymer poly (2,6-dimethyl-1,4-phenylene oxide) (PPO) by bromination and quaternary amination

While for second membrane series, IEC remains approximately unchanged by the aryl substitution but the fixed concentration increases with aryl substitutions due to a decrease in the water content entrapped in the membrane. Therefore, by adjusting the portions of benzyl and aryl substitutions, one can reasonably obtain a membrane with both desired IEC and appropriate water content for acid and salt separation [26-27].

Table 1. Properties of example membrane series developed in this paper[a]

Membrane Number	ABC[b], %	BBC[b], %	IEC (meq/g dry membrane)	W_R (g.H_2O/g dry membrane)	C_R (M)	R_m ($\Omega.cm^2$)
1	0.16	0.116	0.89	0.418	2.13	4.82
2	0.16	0.173	1.27	0.50	2.52	3.81
3	0.16	0.285	1.94	0.593	3.26	3.72
4	0.16	0.310	2.08	0.74	2.80	1.93
5	0.16	0.328	2.18	0.77	2.82	1.93
6	0.10	0.285	1.935	0.673	2.87	1.82
7	0.28	0.285	1.952	0.478	4.08	3.81
8	0.38	0.285	1.943	0.442	4.39	4.32
9	0.42	0.285	1.933	0.387	4.99	4.93
10	0.54	0.285	1.920	0.302	6.36	5.25

a. All the measurements were conducted at temperature of 25°C.

b. ABC: Aryl bromine content, BBC: benzyl bromine content. Both are assumed Mono-substituted

Another example is given for the alteration of membrane properties by in situ-crosslinking with di-amine [25]. Fig. 2a demonstrates the IECs of the membrane simultaneously quaternized and crosslinked with trimethylamine (TMA) and ethylenediamine (EDA) mixtures. Obviously, the IEC increases with tertiary amine content in the amination medium gradually and finally, plateau value is attained for all ethylene diamine compositions. Though IEC slightly decreases with diamine content in the medium, it is compensated by a dramatically increase in volumetric stability as shown in Fig. 2b.

An alternative way is the surface crosslinking with amine aqueous solution [28]. The process can be conducted before or after quaternary amination with TMA. Advantage of this way is that the membrane other properties such as area resistance or water content can be changed with IEC remaining approximately unchanged. So the fixed group concentration, which is a ratio of IEC to water content, can be changed to great extent. As an example, Fig. 3 shows the results of surface crosslinking of anion exchange membrane with 1 M ammonium aqueous solution at different time stages. It is observed that with an increase in crosslinking time, the water content decreases, membrane area resistance increases while the IEC keeps approximated changes, allowing three categories of membranes separating by the dashed line. The first category near the left of the first dashed line show high water content and thus can be used in diffusional dialysis for recovery wastes containing strong acid of high concentration and cations of high valence. The category between the two dashed lines with middle water content and accepted area resistance are expected to use in electrodialysis, polymer electrolyte for fuel cell or diffusional dialysis for recovery wastes containing strong acid of high concentration and metal complex anions.

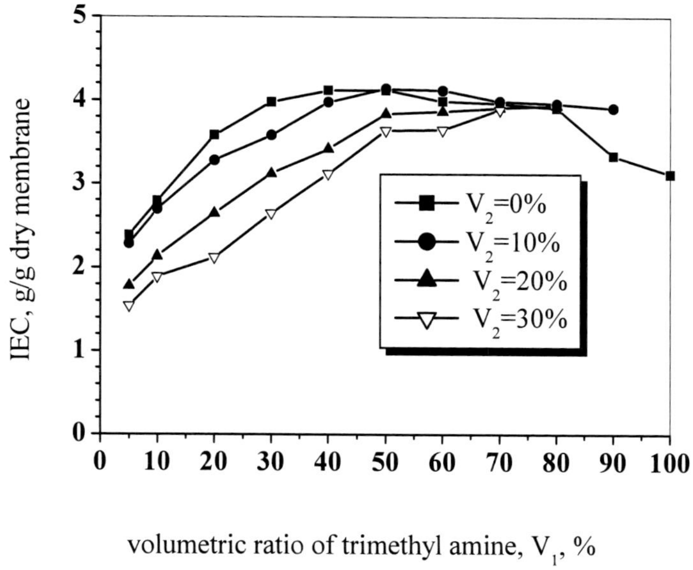

(a)

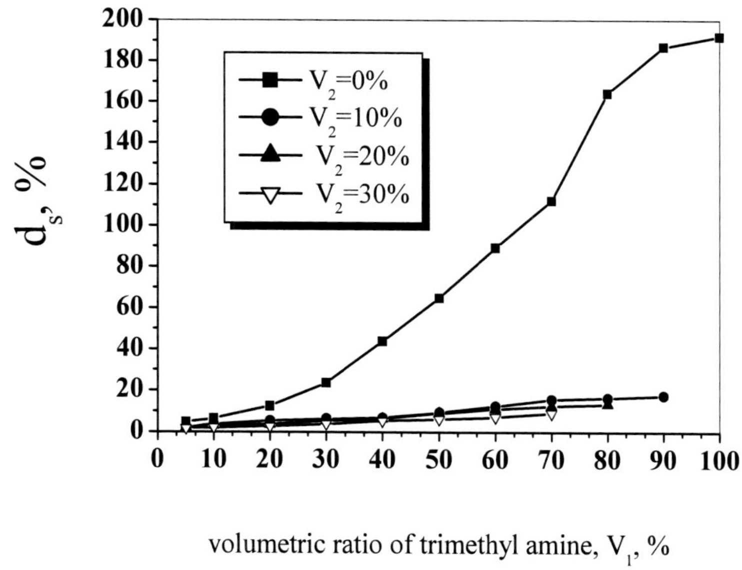

(b)

Fig. 2. Effect of volumetric ratios of trimethyl amine (V_1) and ethylenediamine (V_2) on membrane IEC (a) and membrane dimensional stability (b) with time=10 hrs, temperature 35^0 C Benzyl substitution of PPO=0.992 [25]

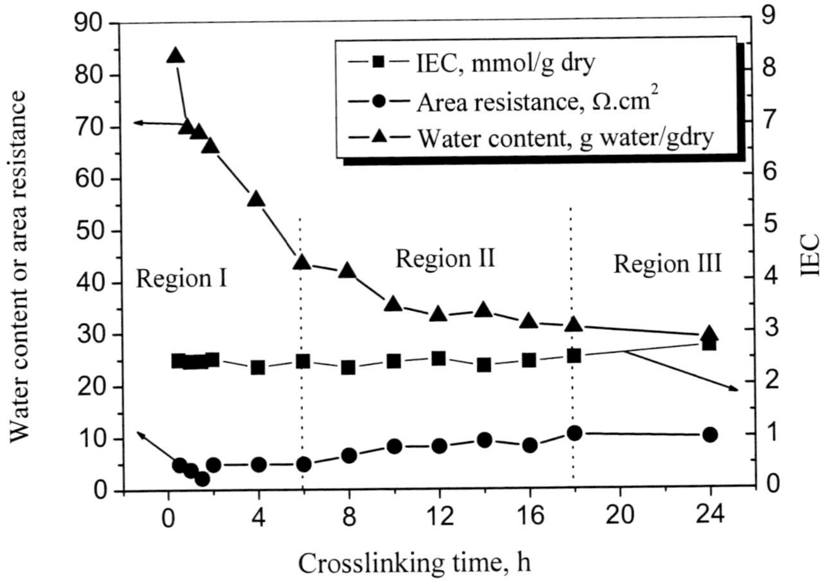

Fig. 3. Effect of surface crosslinking time on the main properties of anion exchange membranes

And the last category near the right with low water content and high area resistance is applicable at the occasions where both the acid recovery ratio and the selectivity are not so important but the purpose is to decrease the acid concentration of waste containing relatively low concentration acid and valuable metal ions and then recycle the waste to recover the metal [28]. Furthermore, due to an introduction in crosslinking structure, the membranes' resistance to chemical solvent has been largely improved as shown in Table 2 which demonstrated the solubility results of the base membranes and the final membranes as well as non-crosslinking anion exchange membranes S_{203} prepared from quaternary animation of chloromethylated polysulfone. It was found that the resultant membrane has extraordinary excellent resistance to most conventional solvents, especially at a relatively high temperature compared with the raw polymer and an analogous commercial membrane S_{203} in China.

The membrane properties can be controlled not only by the bromination and crosslinking process, but also by type of tertiary amines used for quaternary amination. Table 3 showed the main properties of anion exchange layer series with different functional group types [29-31]. It is observed that IEC decreases with the order of functional groups Trimethylbenzylammonium< Dimethyethanolbenzylammonium <Triethylbenzylammonium< Tripropylbenzylammonium< Tributylbenzylammonium, roughly indicating that with an increase in alkyl length of amines, the quaternary amination reaction becomes more difficult. The W_R and R_m follow a reasonable order: W_R decreases in this order while R_m increases in this order. This trend can be interpreted by the intrinsic hydrophilicity of different tertiary amines. In general, the hydrophilicity of compounds is estimated by the hydrophilic lipophilic balance (HLB) values, which are defined as [30]

$$HLB = \sum(\text{hydrophilic group numbers}) - \sum(\text{hydrophobic group numbers}) + 7$$

Table 2. Solubility comparisons between the membrane, raw polymer and S203 membrane*

	PPS		Membrane in this paper		S_{203}	
	25℃	50℃	25℃	50℃	25℃	50℃
Chlorobenzene	+	+	—	—	Δ	Δ
Toluene	+	+	—	—	Δ	Δ
Chloroform	+	+	—	—	Δ	+
Dichloroethane	+	+	—	—	Δ	+
cyclohexane	Δ	+	—	—		
N-Hexane	Δ	+	—	—		
Ethyl ether	—	—	—	—		
Acetone	—	Δ	—	—	—	—
tetrahydrofuran	—	Δ	—	—	—	Δ
Ethyl acetate	—	—	—	—		Δ
Ethanol	—	—	—	—		—
Methanol	—	—	—	—	—	—
n- propanol	—	—	—	—	—	—
Dimethyl sulfoxide	—	Δ	—	—	—	Δ
dioxane	Δ	Δ	—	—	—	—
Dimethyl formamide	—	—	—	Δ	+	+
Dimethyl acetamide	—	—	—	Δ	+	+
Acetic acid	—	—	—	—	—	—
water	—	—	—	—	—	—

*. +, soluble; Δ, partly soluble or swollen to great extent; —, insoluble; concentration is 0.1%

Table 3. Main properties of anion exchange layers with different functional groups type

Amine	Molecular formula	HLB value*	IEC (meq/g dry membrane)	W_R (g.H_2O/g dry membrane)	R_m ($\Omega.cm^2$)
Methyl amine(MA)	NH_2CH_3	15.93	1.60	0.046	70.3
Dimethyl amine(DMA)	$NH(CH_3)_2$	15.45	1.68	0.081	40.4
Trimethylamine(TMA)	$N(CH_3)_3$	14.98	2.22	0.822	3.01
Triethylamine(TEA)	$N(CH_2CH_3)_3$	13.55	0.53	0.079	74.6
Tripropylamine(TPA)	$N(CH_2CH_2CH_3)_3$	12.13	0.27	0.054	100.1
Tributylamine(TBA)	$N(CH_2CH_2CH_2CH_3)_3$	10.7	0.23	0.033	270.3
Dimethyethanolamine(DMEA)	$(CH_3)_2NCH_2CH_2OH$	16.40	0.78	0.087	52.3

*Calculated from HLB groups numbers: N (primary, secondary and tertiary amine) 9.4; $-CH_2-$ 0.475; $-CH_3$ 0.475;-OH 1.9.

HLB values shown in Table 3 are calculated from HLB groups numbers of amines that were bonded to benzyl of BPPO as charged groups. Obviously, hydrophilicity of the anion exchange layers shows consistency with HLB values of amines by measuring the water content of the layers except the case that water content of anion exchange layer with dimethyethanolaminium group is less than that with trimethylbenzylammonium group though the HLB value of dimethyethanolamine(DMEA) is higher than that of TMA. This is mainly due to the steric effect of amination reaction of DMEA because the IEC of the former is far less than the latter.

Bromination is a simple process and the obtained polymers can be also further modified to permit multi-functional anion exchange membranes. For example, the thermo-sensitive membranes can be prepared by ^{60}Co-γ-rays irradiation-induced simultaneous grafting N-isopropylacrylamide (NIPAAm) onto btominated (BPPO) membrane [33-34]. Fig. 4 showed thermo-sensitive water permeation flux of grafted membrane (grafting degree=7.27%) compared with substrate membrane at the operational pressure 0.1MPa [33] and Fig. 5 demonstrated the thermally controlled release properties for ionic drugs sodium salicylate through the thermal anion exchange membrane prepared by quaternary amination of the thermal sensitive base membrane shown in scheme 1(g) [34]

In summary, by simple bromination, the hazardous CME was avoided and the obtained polymers can be further treated by amination of different amines or pyridine, crosslingking with diamine or ammonium aqueous solution, grafting copolymerization of themo-sensitive monomer such as N-isopropylacrylamide (NIPAAm), etc., to give series of multi-functional membranes. Scheme 1 summarized the reaction processes as well as the structure of the obtainable membranes and Fig. 6 schematically demonstrated the structure and function control of the membranes as well as their possible applications.

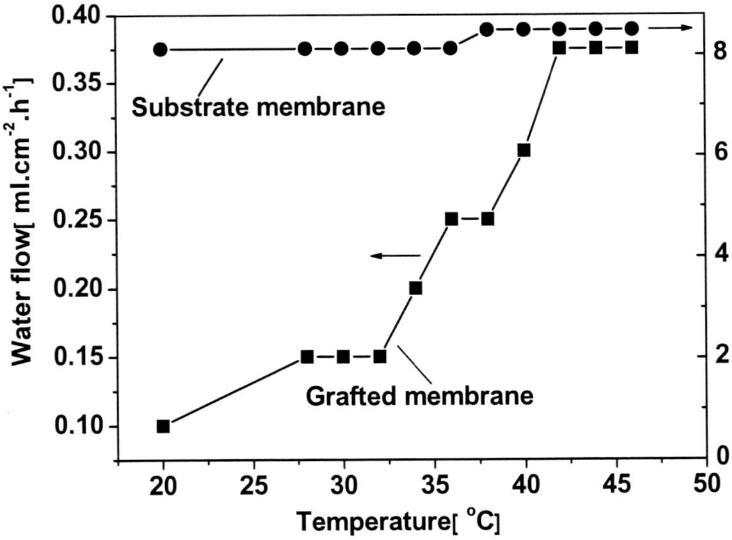

Fig. 4. Comparison of thermo-sensitive water permeation flux of grafted membrane (grafting degree=7.27%) with that of substrate membrane at the operational pressure 0.1Mpa [33]

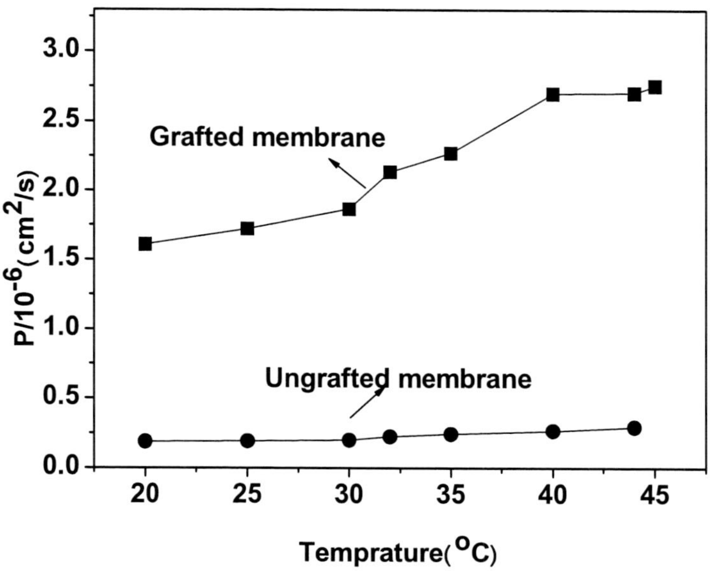

Fig. 5. The permeability coefficients of sodium salicylate across the grafted charged membrane (IEC=0.400mmol/g) and the ungrafted charged membrane (IEC=0.673mmol/g) in response to temperature change between 20 °C and 45 °C in the deionized water [35]

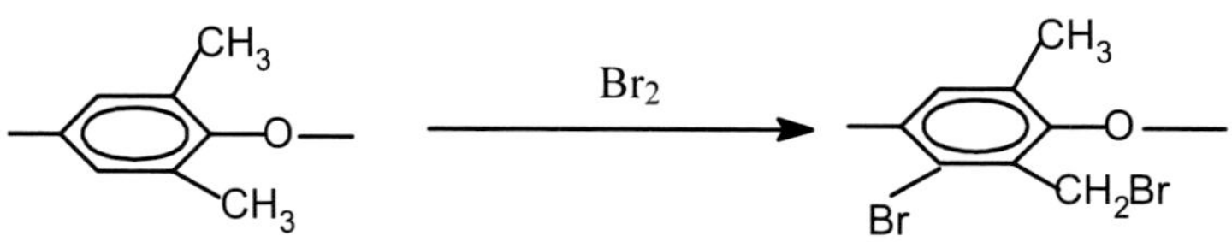

(a) Bromination

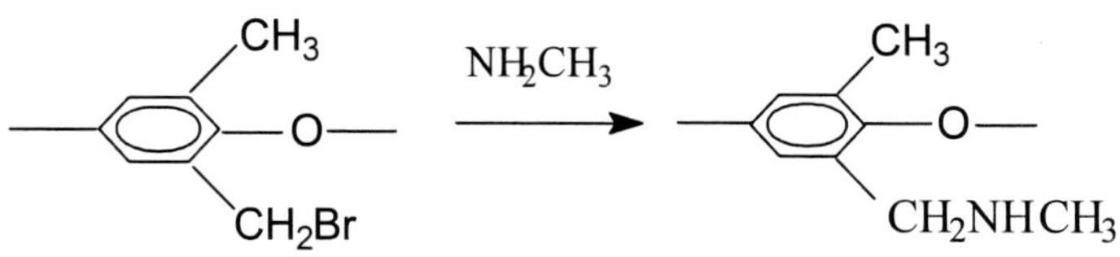

(b) Secondary ammonium group

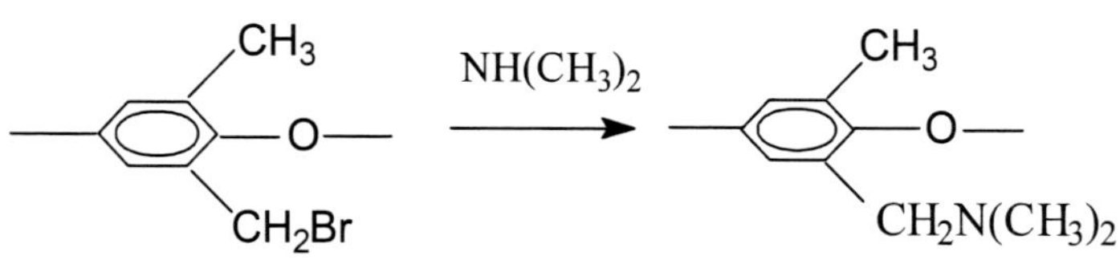

(c) Tertiary ammonium group

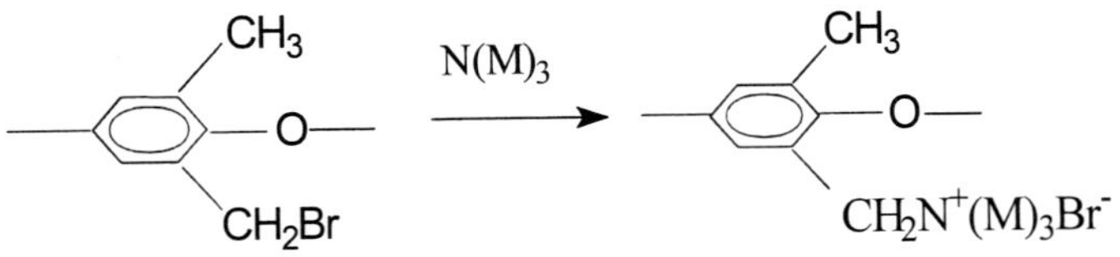

M=CH_3, CH_3CH_2, $CH_3CH_2CH_2$, $CH_3CH_2CH_2CH_2$

(d) I type quaternary ammonium group

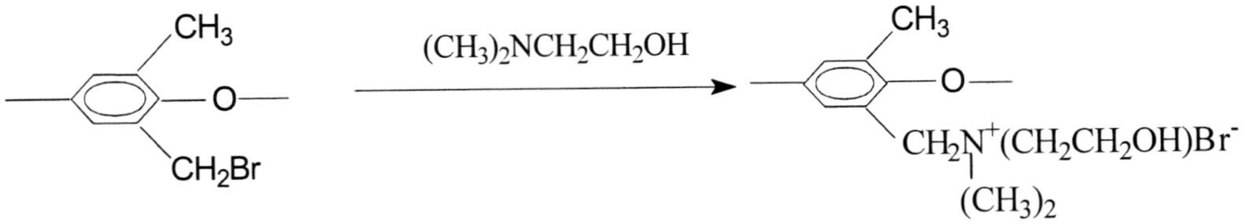

(e) II type quaternary ammonium group

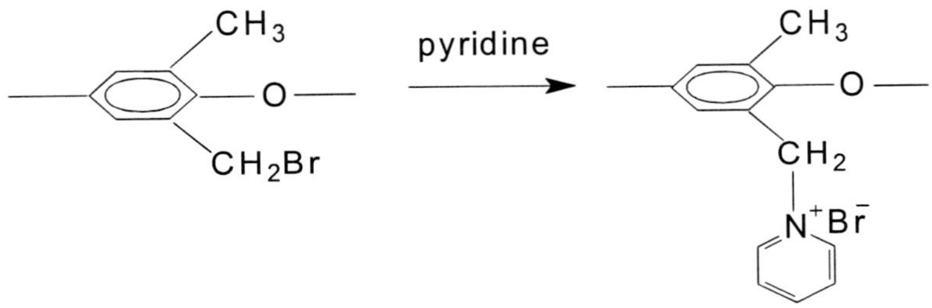

(f) Typical less hydrophilic membranes with pyridine groups

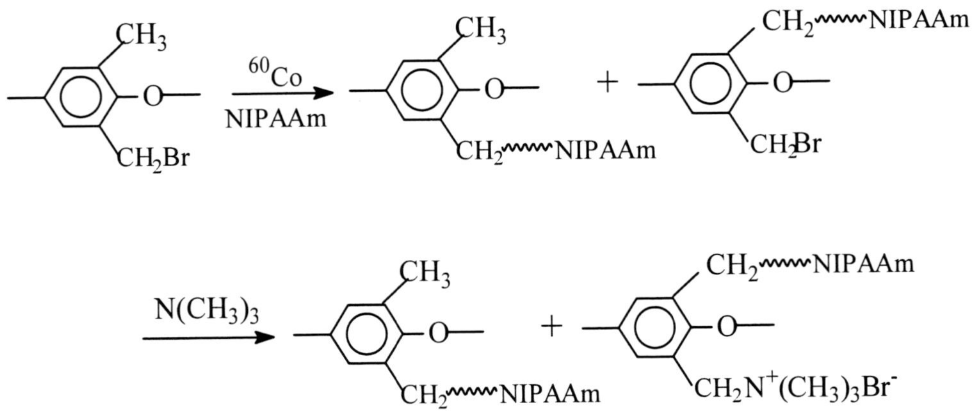

(g) Thermo-sensitive anion exchange membrane by Radiation-induced grafting of N-isopropylacrylamide (NIPAAm)

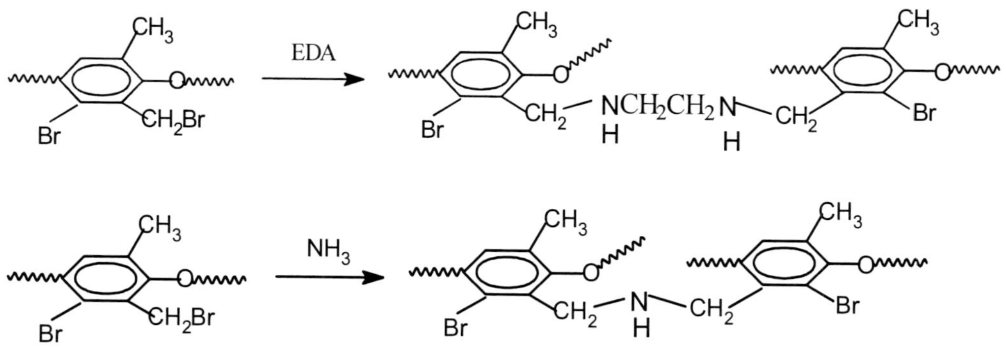

(h) Typical crosslinking with di-amine or ammonium aqueous solution to give crosslinking membrane

Scheme 1. The reaction and the structure of series of anion exchange membranes prepared from bromomethylated poly(2,6-dimethyl-1,4-phenylene oxide) (BPPO).

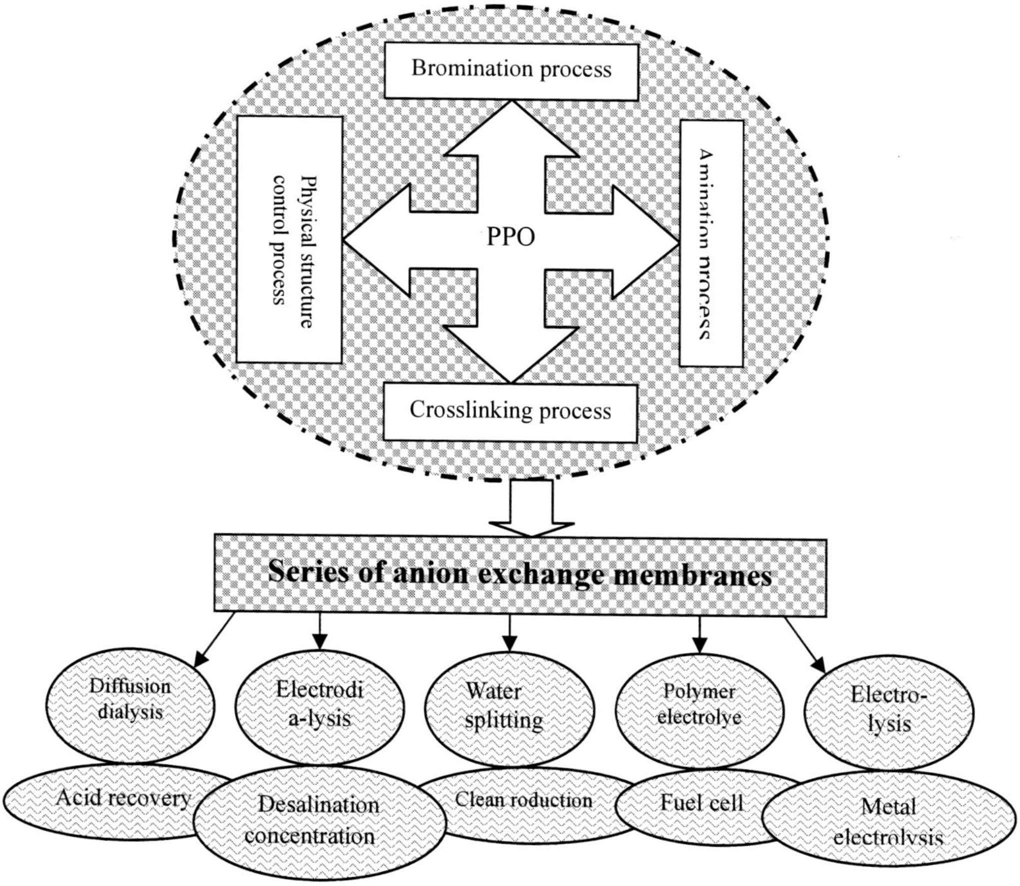

Fig. 6. Schematically presentation of series of anion exchange membranes developed from poly (2,6-dimethyl-1,4-phenylene oxide) (PPO) for various electro-membrane processes.

3. Anion Exchange Membranes Preparation by Chloroacetylation Reaction (Friedel-Crafts Reaction)

The bromination provided a convenient and economic method for anion exchange membranes without using hazardous materials chloromethyl methyl ether (CME) and bis-chloromethylether (BCME). However, such method is limited by the structure of the base polymers, i.e., it is requested that the aromatic polymers possess the side methyl group attached to the benzyl ring.

Chloroacetylation of aromatic polymers provides a easy way for anion exchangers to avoid the additional crosslinking of the chloromethylated or bromomethylated polymers. It is not limted by the structure of the aromatic polymers (side methyl group is not definitely needed). This method has been used to prepare the anion exchange resins by Xu and Hu based on crosslinked polystyrene.[35] As shown in Scheme 2, the processes comprise first introduction of acetyl groups rather than chloromethyl groups as substituents for aromatic nuclei in the copolymer, by treatment of the copolymer with acetyl chloride or acetic anhydride in the presence of a Friedel–Crafts catalyst, then introduction of basic groups into the copolymer by reductive amination [36], i.e. treatment of the acetylated intermediate with ammonia or amines in the presence of a reducing agent. The resulting resins, if desired, can be readily converted to quaternary ammonium, strong base anion exchangers by reacting with methyl iodide (CH_3I) or dimethyl sulfate. Such process can prepare either weak base anion exchangers or strong base anion exchangers. Furthermore, the chemical structures of the obtained anion exchangers is a little different from that of conventional resins, that is, it contain an additional side-methyl on the obtained structure of $-CH(CH_3)N<$ in comparison with typical structure $-CH_2N<$. The water content and ion exchange capacity has been reported in Table 4 [35] for different functional groups.

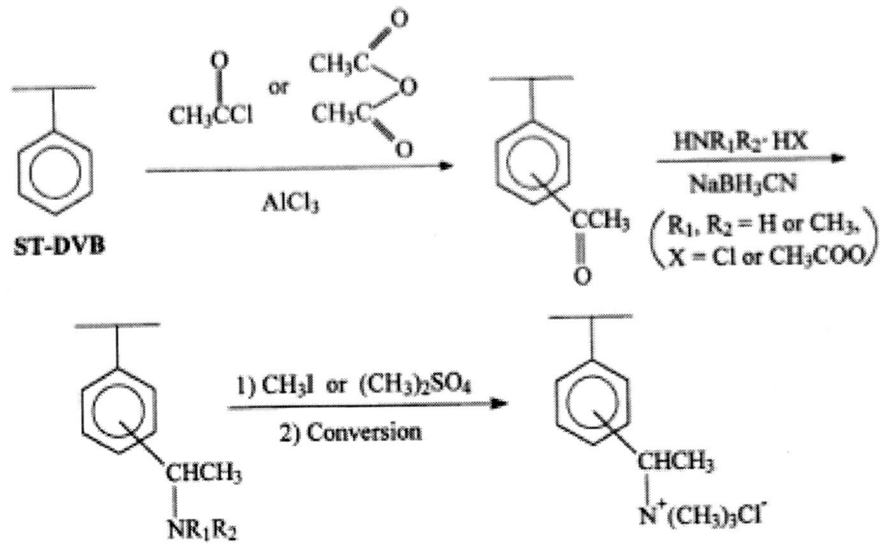

Scheme 2. The new synthetic route for anion exchangers based on crosslinked polystyrene, taken from [35]

Table 4. Water content and IEC for the anion exchangers of different group reported by Xu and Hu [35]

Type of resins	Form	Crosslinking	Water content (%)	Exchange capacity(mmol/g)	
				Calculated	Experimental
Primary amino	Free base	8	33.3	3.34	2.95
Secondary amino	Free base	20	18.5	3.17	3.05
Tertiary amino	Free base	8	23.3	0.59	0.53
Quaternary ammonium	Chloride	8	58.9	3.25	3.10

The route in Scheme 2 can also be modified by first acetylation of styrene-divinylbenzene copolymers by Friedel - Crafts reaction followed by bromination and amination as shown in Scheme 3 [37]. The obtained resins has much common in intrinsic properties with resins prepared by Schme 2. Though Schemes 2-3 are for anion exchange resins, they can be also lent to prepare anion exchange membranes, because the difference between anion exchange resins and membranes are only in the physical form.

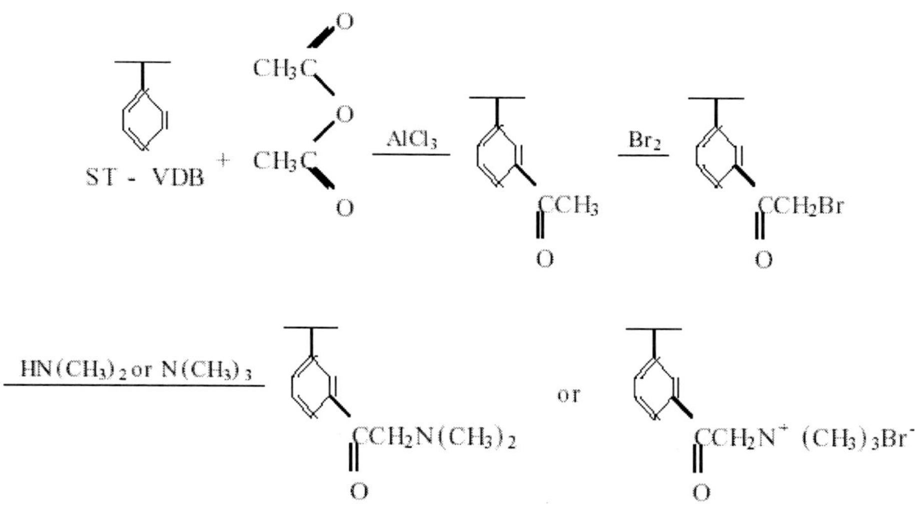

Scheme 3. Preparation of anion exchangers based on crosslinked polystyrene by first acetylation, bromination and amination, taken from [37]

To find the new route free of hazardous materials and also continue the preparation of anion exchange membranes from the linear engineering plastic poly (2,6-dimethyl-1,4-phenylene oxide) (PPO), recently, the above method for anion exchanger was used to prepare anion exchange membrane in our lab. PPO was treated with 1,3-bis-chloroacetyl in the presence of a Friedel-Crafts catalyst to obtain chloroacetyl groups rather than chloromethyl groups as substituents for aromatic nuclei in the polymer, then the chloroacetylated PPO (CPPO) was converted into quaternary ammonium in the amination reagent [38]. The whole reaction process is shown in Scheme 4. Chloroacetylation process was largely influenced by the quantity of the catalyst (anhydrous $AlCl_3$) and the reaction temperature. As shown in Table 5, at a certain temperature, the substitution degree grows as the quantity of the catalyst increases in the cases of whether the equal molar ratio of PPO to CAC (c.f. Nos. 1-3 in Table 5) or that of CAC to $AlCl_3$ (c.f. No.s 4-6 in Table 5).

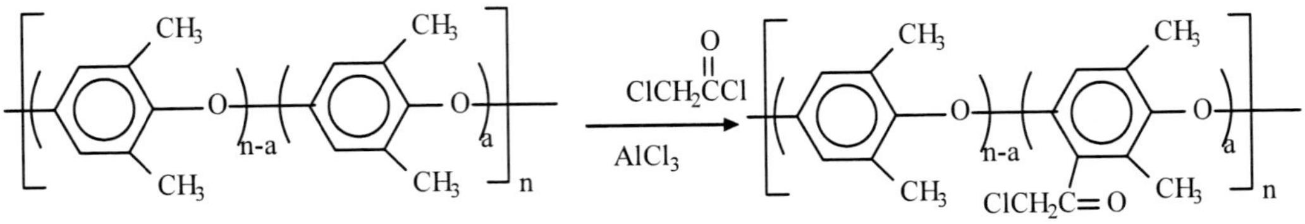

Step 1. Chloroacetylation

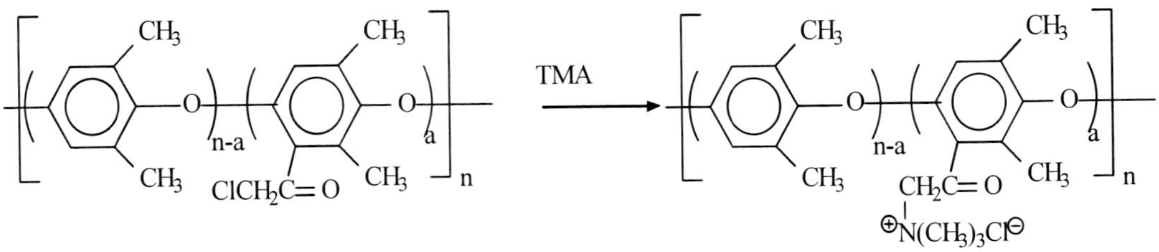

Step 2. Quaternary amination

Scheme 4. The new synthetic route for anion exchange membranes based on PPO

Table 5. Chloroacetylation of PPO under Friedel-Crafts Condition

Code number	PPO:CAC[a]:$AlCl_3$ (mol:mol:mol)	Temp (℃)	Substitution degree[b] (%)	Naked observation
1	1.00:1.00:0.50	40	22.3	cracked
2	1.00:1.00:0.75	40	31.2	cracked
3	1.00:1.00:1.25	40	63.2	good
4	1.00:0.50:0.50	40	32.9	cracked
5	1.00:0.75:0.75	40	33.2	cracked
6	1.00:1.50:1.50	40	71.8	fragile
7	1.00:1.00:1.00	20	37.5	cracked
8	1.00:1.00:1.00	30	44.9	Excellent
9	1.00:1.00:1.00	40	50.3	Excellent
10	1.00:1.00:1.00	50	70.6	fragile
11	1.00:1.00:1.00	60	83.5	fragile

[a]chloroacetyl chlorid; [b]determined by ^{1}H NMR;

At a given ratio of reactants and catalyst(Nos. 7-11 in Table 5), the substitution degree increases appreciably with reaction temperature. Unlike the anion exchange resins, not all the substitution degrees are suitable for anion exchange membranes preparation. The results of membrane formation were qualitatively evaluated by the naked eyes and shown in the last column of Table 5. It was found that the substituted polymers with chloroacetylation degree around 50% were easy to form homogeneous base membranes, while the polymers with higher or lower chloroacetylation degree were not suitable for membrane formation. The products with low substitution degree contain so large amount of unreacted PPO that it can not dissolve in the solvents such as DMF to form base membrane (No. 1, 2, 5 in Table 5). On the contrary, product with high substitution degree is obtained under acute reaction conditions, such as high temperature, adding more catalyst, etc., and this may lead to a destroy of the pliancy of resultant polymer chain in a certain extent and the formed membrane is very fragile (No.3,6,10,11 in Table 5). Upon the determination of substitution degree, the intrinsic properties of the membranes will be largely dependent on the amination process such as amination time, temperature and concentration of TMA. As examples, Table 6 illustrate the prepared anion exchange membrane aminated with different concentration of TMA aqueous solutions ranging from 0.7 to 2.5 mol/l for 48h and the temperature is fixed at 30°C. when compared with anion exchange membranes prepared from bromomethylated PPO, the conversion ratio for the quaternary amination reaction is relatively low , giving rise to low IEC values and water content and higher area resistance. Therefore, this comparison demonstrated that the quaternary amination for bromomethylated PPO is quicker than that for CPPO due to the steric hindrance of cloroacetylated groups.

Table 6. The intrinsic properties of anion exchange membranes prepared by chloroacetylation of PPO followed quaternary amination at various concentratioon of TMA. (substitution degree=50.3%, amination temperature =30 °C, amination time=48h)

Code number	Concentration of TMA, M	Water uptake (g. water /g wet membrane)	IEC, mmol/g dry membrane	Area resistance, $\Omega.cm^2$
1	0.77	0.619	0.76	188
2	0.91	0.637	0.78	170
3	1.12	0.656	0.8	89.2
4	1.43	0.656	1.03	34.4
5	1.67	0.67	1.17	18.6
6	2.5	0.678	1.16	18.7

4. Anion Exchange Membranes Preparation by Copolymerization of Epoxy Group Containing Monomers

As well known, substance with epoxy groups are very active and tend to react with amines to form anion exchange groups. So such route can often be used to synthesize anion exchange membranes without using the hazardous materials CME or BCME. To increase the mechanical strength, epoxy acrylate monomers such as glycidyl methacrylate or glycidyl ethacrylate are often copolymerized with other monomers such as divinylbenzene (DVB) to produce the base membrane and then subsequently aminated with triethyl amine to prepare anion exchange membranes. Scheme 5 presents a reaction route for preparing such anion exchange membrane [39]. The main intrinsic properties of the obtainable membrane are as follows: IEC=1.8-2.2 mmol/g dry membrane, water content =23%~33% [39].

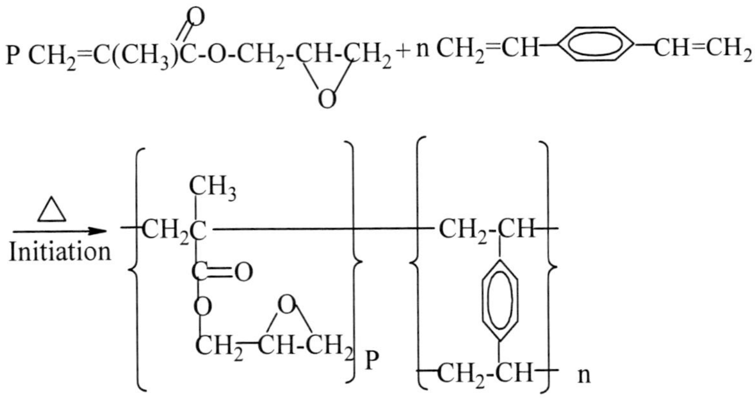

Step 1. Copolymerization reaction

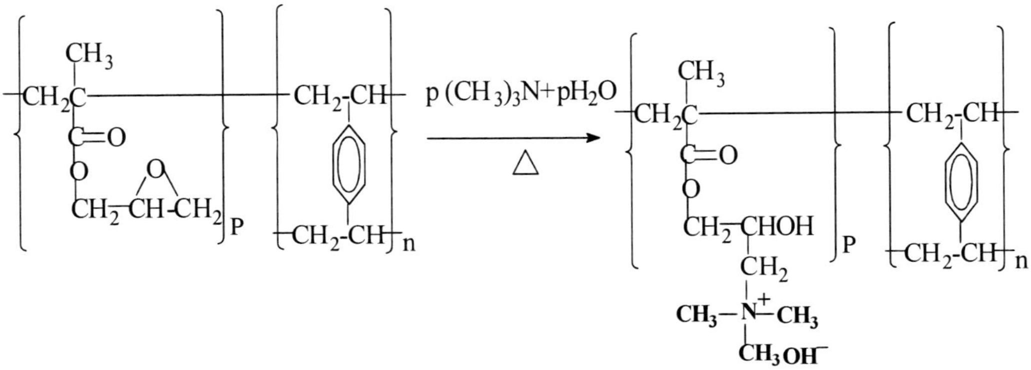

Step 2. Quaternary amination reaction

Scheme 5. Anion exchange membrane preparation by ring opening reaction of epoxy group instead of chloromethylation with chloromethyl methyl ether

Bedsides the directly copolymerization into base membrane, epoxy acrylate monomers and divinylbenzene (DVB) can also conduct to graft polymerization onto a inert polymeric film. For example, K. Sunaga et. al. prepared porous anion-exchange membranes of a flat-sheet form through co-grafting of an epoxy group containing monomer, glycidyl methacrylate (GMA) and a cross-linker, divinylbenzene (DVB), onto a porous polyethylene membrane, followed by the introduction of a diethylamino (DEA) group [40] . The remaining epoxy groups of the G/D-AE and G-AE membranes were reacted with ethanolamine. The whole process was schematically shown in Scheme 6. The results demonstrated that a monomer mixture of GMA and DVB containing 1-10 mol % of DVB provided a DVB-cross-linked graft chain ranging from 15 to 45 mol % of DVB. After the introduction of the DEA group, the swelling of the membrane in water was reduced and protein binding capacity increased due to a hydrophobic interaction [40].

Polymer with epoxy groups can be used to synthesize the anion exchange membranes. Such an example is the traditional commercial anion exchange membranes made from polyepichlorohydrin rubber in China. As shown in Scheme 7, in the presence of triisobutyl aluminium, chloroepoxy propane monomers was polymerized onto a support and then quaternized with TMA to obtain such anion exchange membrane whose IEC can attain as high as 1.2-1.6 mmol/g dry membrane.

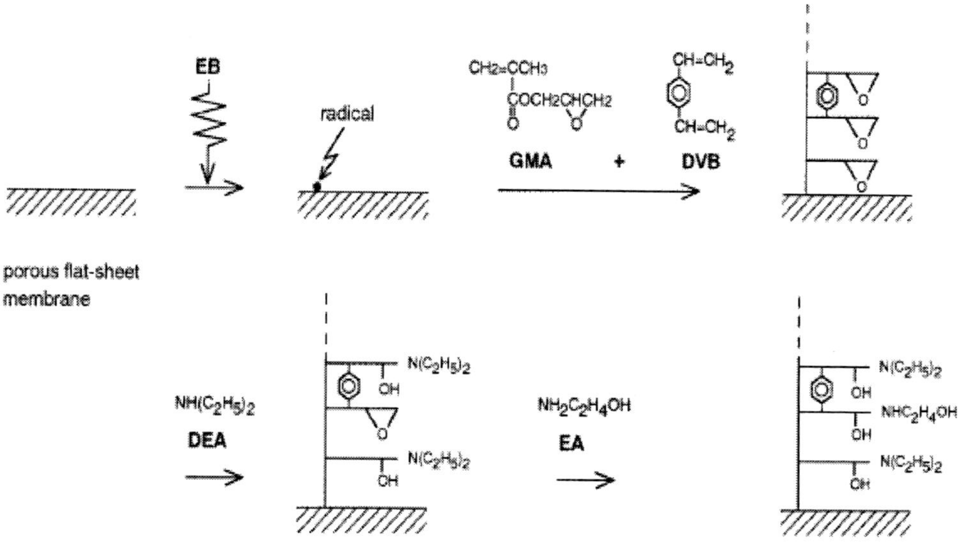

Scheme 6. Preparation scheme for a porous anion-exchange membrane by radiation-induced cografting of GMA with DVB, and subsequent chemical modifications, taken from [40].

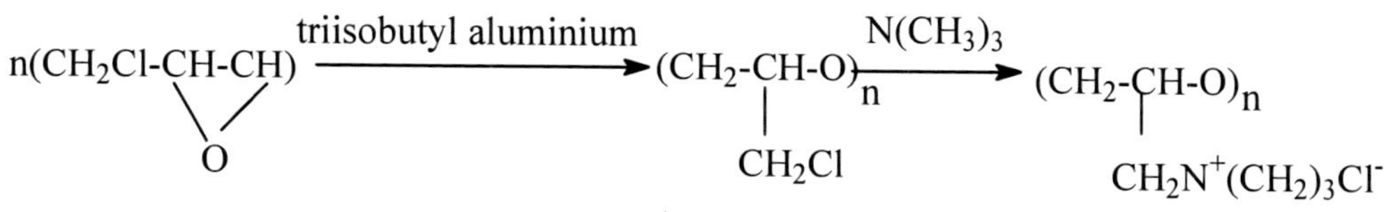

Scheme 7. Anion exchange membranes preparation from polymerizations of chloroepoxy propane monomers

5. Anion Exchange Membranes Preparation by Long Chain Haloalkylethers

As mentioned previously, the chloromethylation of aromatic polymers has been a key step in the preparation of anion exchange membranes. Chloromethyl methyl ether(CME) and bis(Chloromethyl)ether (BCME) have been very successfully applied in the chloromethylation reactions, providing excellent conversion and yields,. In the early seventies, CME, BCME and their bromine analogues were reported to be notorious carcinogens, which caused phasing-out of these agents [14-15]. Since then, special attention has been given to the alternative reagents. Among of them, halomethylation with BCME free long chain haloalkylethers have been approved to be safe and used extensively [41-44]. For example, Warshawsky and coworkers [42-44] have introduced chloromethyl (or bromomethyl) octyl ether (CMOE or BMOE) as the halomethylating reagent. These long chain halomethylalkyl ethers free of BCME or CME can be synthesized by the reaction of paraformaldehyde with alkanol and HX(X=Cl or Br) gas in an organic solvent, preferably a halogenated hydrocarbon, e.g. CH_2Cl_2, $CHCl_3$, 1,2-dichloroethane(DCE) or 1,1,2-

trichloroethane(TCE) according Scheme 8(a)[41-42] The obtainable long chain halomethylalkyl ethers $ROCH_2X$ were applied in the halomethylation of aromatic substrates and polymers according to Scheme 8(b).

Even through it was reported that sometimes, the synthesis of halomethylalkyl ethers shown in Scheme 8a can lead to side products bis(Chloromethyl)ether (BCME) or bis(bromomethyl)ether (BBME), the quantity can be controlled by selecting a primary alkanol and controlling the ratio of paraformaldehyde to ROH [41-42]. It was laso shown that when polymer polysulfone was halomethylated in this way, the modified polymers preserves the excellent film-forming properties and is , furthermore, compatible with unmodified polysulfone. The main properties of the obtainable anion exchange membranes prepared with various degrees of substitution were listed in Table 7 in comparison with those prepared with various mixtures.

$$ROH+(CH_2O)_n \xrightarrow[S/0\text{-}5^{\circ}C]{HX} ROCH_2X$$

X=Cl,Br; Conc.=3M
S=1,2-DCE,$CHCl_3$, 1,1,2-TCE

(a)Preparation of halomethylalkyl ethers

$$ROCH_2X+ArH \xrightarrow[s/r.t.]{SnCl_4} ROH+ArCH_2X$$

(b) Halomethylation

Scheme 8. Preparation of halomethylalkyl ethers and halomethylation of aromatic polymers

Table 7. Main properties of anion exchange membrane prepared by mixtures of bromomethylated polysulfone (BMPS) and polysulfone [44]

IEC mmol/g BMPS	IEC mmol/g mixture	PS, Wt.%	Transport number 1N/2N	Transport number 0.1N/0.2N	Area resistance , $\Omega.cm^2$
1.08	1.08	0	0.82	0.99	16
1.39	1.08	22.5	0.83	0.90	6.6
1.70	1.08	36.3	0.83	0.94	7.6
1.39	1.27	8.8	0.82	0.92	4.0
1.70	1.27	25	0.81	0.94	4.7
1.70	1.39	17.87	0.81	0.88	2.6
1.70	1.70	0	0.77	0.90	1.7

6. Anion Exchange Membranes Preparation by Sol-Gel Route

Sol–gel process makes possible a relatively easy incorporation of pure inorganic phase into an organic matrix. A low temperature procedure, high purity of reactants and possibility of mixing of precursors of organic and inorganic phases on molecular level are the main advantage of this technique [45]. Basically, sols are dispersions of colloidal particles (size 1–100 nm) in a liquid and a gel is an interconnected, rigid network with pores of sub micrometer dimensions and polymeric chains whose average length is greater than a micron. Originally, the method has been used to prepare nanostructured organic–inorganic composite materials [45], but currently, it has been used to prepare inorganic-organic (hybrid) ion exchange membranes, especially anion exchange membrane (or positively charged membranes) [46-53]. Due to the composition of inorganic and organic components, it endows the versatility of membrane materials and also CME can be avoided. For example, Kogure et al. prepared a hybrid anion exchange membrane by sol-gel method (metal alkoxide method) or liquid coupling process of silane coupling agents [46]. Through coupling with these agents to silanol group on the surface of silica membranes, the hybrid membranes become insoluble. Further, the inorganic membranes can give enough mechanical strength to the hybrids [46]. The steps of preparation are indicated elsewhere [46] and the coupling mechanism of silane coupling agent is shown in Scheme 9.

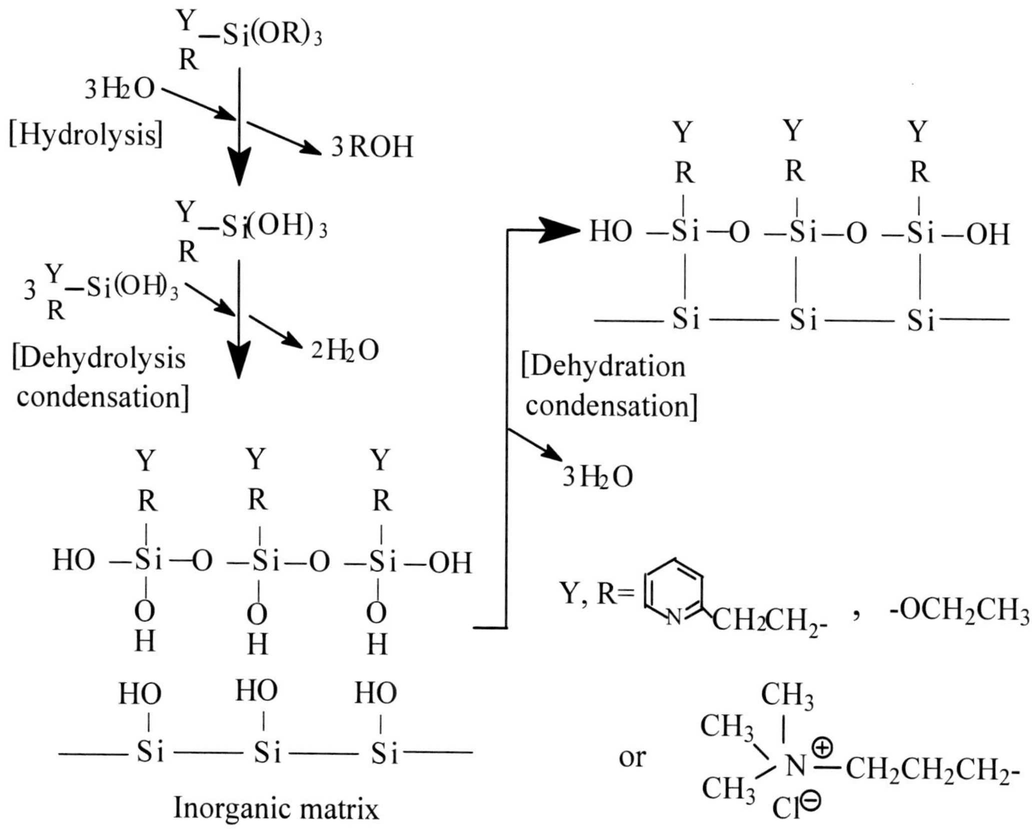

Sceme 9. The coupling mechanism of silane coupling agent [46]

Organic–inorganic composite anion-exchange membranes based on PVA–SiO_2 were also prepared by aqueous dispersion polymerization (sol–gel) method and anion selectivity was introduced in the membrane matrix by the chemical grafting of 4-vinyl pyridine (4-VP) with desired content [47]. Physicochemical and electrochemical properties of the obtainable membranes were highly dependent on the 4-vinylpyridine (4-VP) content in the membrane phase and a better selectivity and membrane conductivity of anion-exchange membranes complete optimization of the loading of 4-VP in the membrane phase is necessary. In all these composite membranes, inorganic content is fixed, which contributes towards the reduction of swelling properties of PVA, while 4-VP, which is essential for the introduction of ionogenic functional groups contributes towards the increase in the swelling properties of the membrane matrix and a membrane with 25% loading with 4-VP exhibited very good selectivity, water content, and ion-exchange capacity along with moderate membrane conductivity, which may be used for their application in electro-driven separation at elevated temperatures or for other electrochemical processes.

Since 2002, preparation of anion exchange membranes by sol-gel has become one of the major jobs conducted in the author's lab. By incorporating the hybrid material idea into the membranes, various routes for hybrid anion exchange membranes were initiated and summarized as the follows.

1) The first route for a hybrid (inorganic-organic) anion exchange membrane was prepared through sol-gel process of trimethoxysilyl functionalized PEO-400, i.e., PEO-$[Si(OCH_3)_3]_2$ and quateramination with ethyl bromide (C_2H_5Br) thereafter. The PEO-$[Si(OCH_3)_3]_2$ was obtained by endcapping polyethylene oxide (PEO)-400 with Tolylene 2,4-diisocyanate (TDI), followed by a reaction with N-[3-(Trimethoxysilyl) propyl] ethylene diamine (A-1120) as shown in Scheme 10 [48]. The anion exchange capacity (IEC) was in the range of $1.6\text{-}3.9\times10^{-2}$ meq.cm^{-2}.

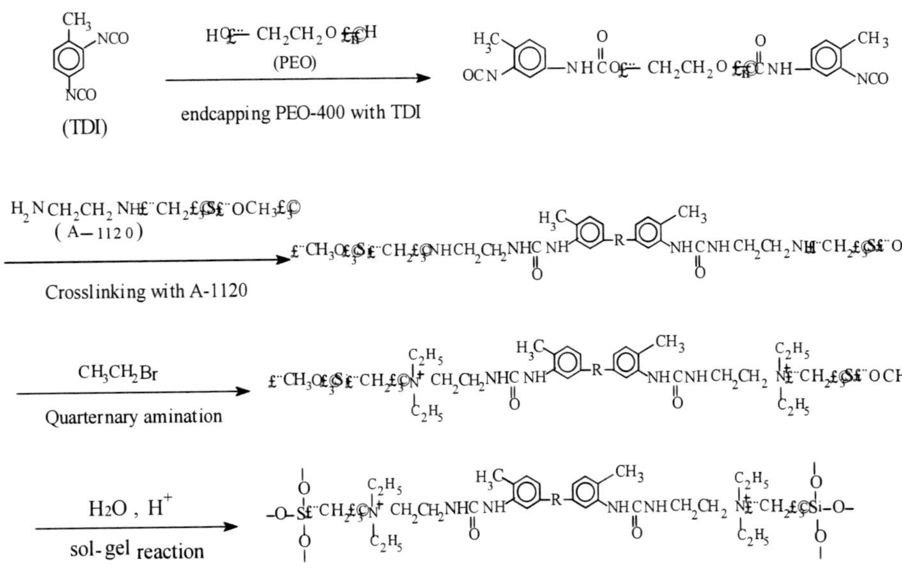

Scheme 10. Preparation of PEO-[Si $(OCH_3)_3]_2$ hybrid anion exchange membrane by sol-gel,R= -NHCOO$(CH_2CH_2O)_n$CONH-

2) An alternative route to a hybrid anion exchange membrane has been prepared from positively charged PMA-SiO_2 nanocomposites, which was synthesized by the sol-gel process of positively charged alkoxysilane-containing polymer precursors. The precursors were synthesized by coupling different amounts of N-[3-(Trimethoxysilyl) propyl] ethylene diamine (A-1120) to poly (methyl acrylate) (PMA), followed by a quaternization reaction and then hydrolysis and condensation as shown in Scheme 11 [49]. As shown in Table 8, anion-exchange capacities of these nanocomposites were shown to be in the range of 0.19-1.20 mmol/g, increasing with the A-1120 content. The low glass transition temperature ensures the elasticity of the hybrids, while the high Td endows the thermal stability.

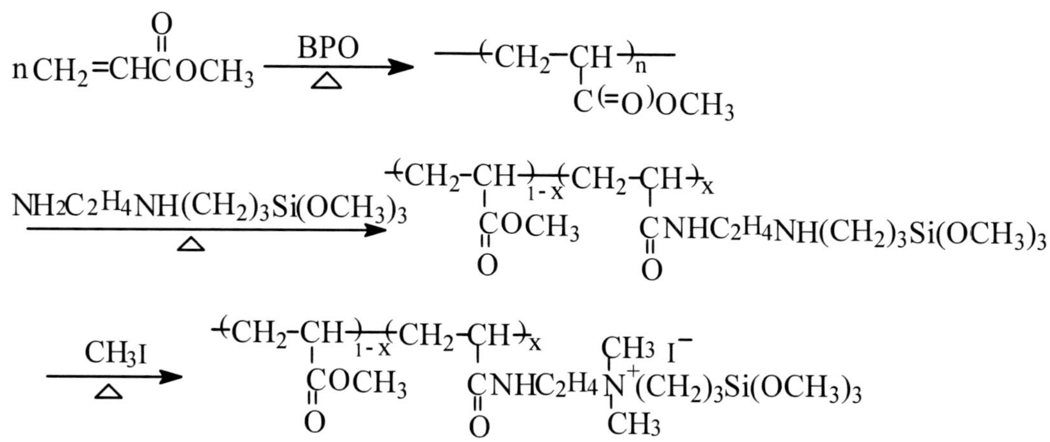

Scheme 11. Preparation of positively charged PMA-A1120 hybrid material and its membrane

Table 8. Thermal properties and anion-exchange capacities(IEc) of PMA-SiO_2 hybrids[126]

	Tg (°C)	Td (°C)	1000 °C residue (wt%)^d	practical IEC (meq/g)	theoretical IEC (meq/g)
PMA-SiO_2 A	16.8	215	8.4	0.19	0.47
PMA-SiO_2 B	23.4	196	9.4	0.52	1.05
PMA-SiO_2 C	20.2	193	12.8	0.56	1.17
PMA-SiO_2 D	31.5	198	14.0	1.20	2.19

3) As described above, modification of PPO is well conducted in our lab for homogeneous ion exchange membrane. To enhance the thermal properties, our third route for a novel hybrid anion exchange membrane was started from bromomethylated PPO by sol-gel, a polymer that has been extensively modified for preparation of polymeric anion exchange membranes in our lab. As shown in Scheme 12, the new hybrid membrane was obtained by introducing 3-aminopropyl-trimethoxysilane (A1110), together with trimethylamine (TMA) into bromomethylated PPO and then forming silica networks by sol-gel process through further condensation and hydrolysis reactions with A1110 [50-51]. The main properties of the hybrids were listed in Table 9. The membranes contained both weak

base and strong base groups and with total IEC values ranging 2.5-3 mmol/g dry. The pure fluxes of the membranes also varied with the compositions in a range of 5-12.2 L/(m^2 bar h). Such membrane can be expected to apply in some circumstance of high temperature because the TGA curves indicated that first decomposition temperature occur at more than 300°C [51].

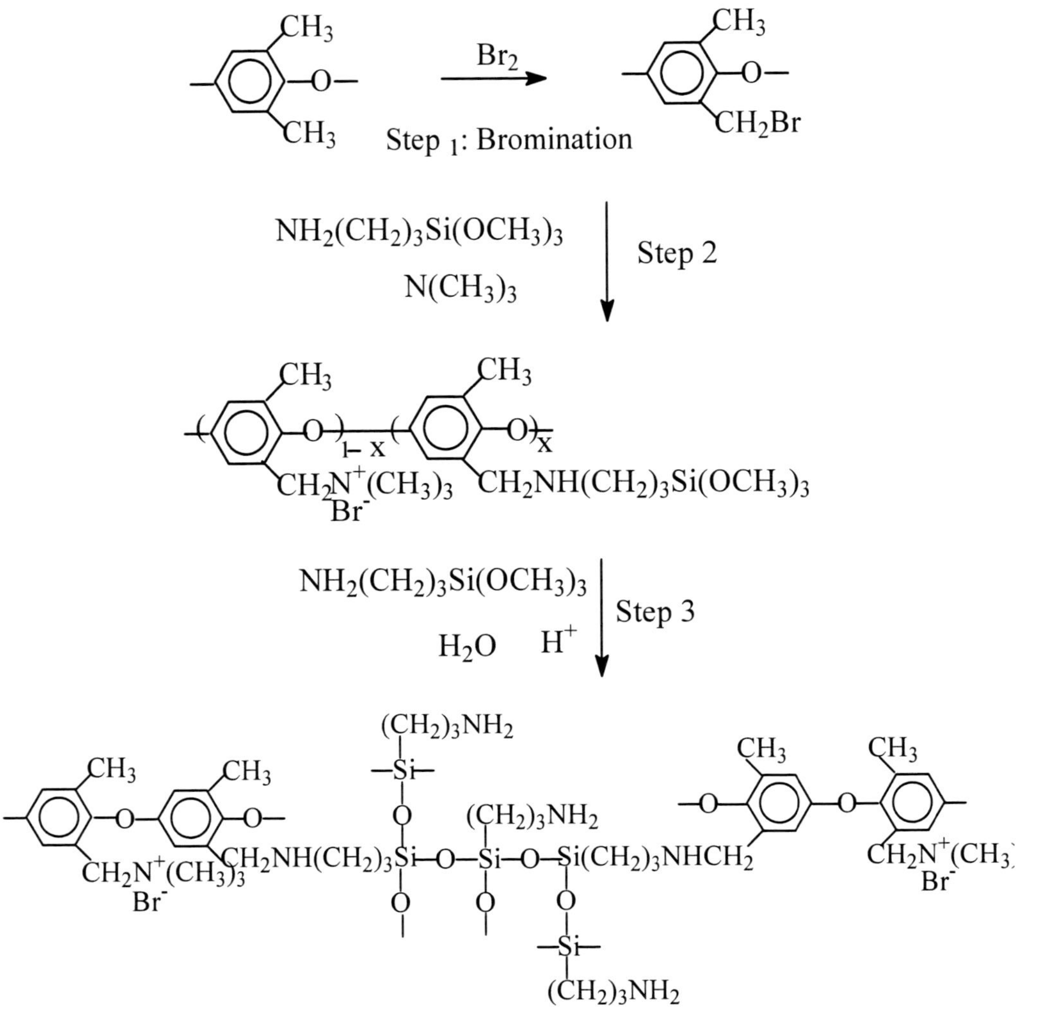

Scheme 12. Preparation of organic-inorganic hybrid anion exchange membrane (materials) based on bromomethylated PPO by sol-gel process [50]

Table 9. Feed composition and the residue in TGA for both the polymer precursors and the prepared membranes

Membranes	Weight ratio of TMA η [a] in feed composition	Residue of polymer precursors in TGA (wt%) [b]	Residue of the prepared membrane in TGA (wt%) [b]	IEC_{total} mmol/g	Water fluxes $L/(m^2\ bar\ h)$
A	0.13	13.00	19.72	2.95	12.26
B	0.07	14.80	20.99	2.73	9.54
C	0.05	16.62	21.78	2.62	9.07
D	0.03	18.47	22.54	2.49	5.09

[a] η = [weight of TMA]/[weight of A1110+ weight of BPPO+weight of TMA]

[b] Residual contents are obtained at 800℃ from TGA thermograms, which were performed at heating rate 10/min under air atmosphere

4) By incorporating the idea of epoxy opening reaction mentioned in section 4, new hybrid anion exchange membranes were prepared through the copolymerization of glycidylmethacrylate (GMA) and γ-methacryloxypropyl trimethoxy silane (γ-MPS) monomers, followed by ring-opening of the GMA moiety with trimethylamine hydrochloride and sol-gel reaction of the γ-MPS moiety [52] (Scheme 13). Compared with the above routes, in which the introduction of the inorganic silica content was through the coupling of organically modified alkoxysilane such as *N*-[3-(trimethoxysilyl) propyl] ethylene diamine (A-1120), phenylaminomethyl triethoxysilane and 3-aminopropyl- trimethoxy silane to the matrix of organic polymer chains ,etc., followed by sol-gel process, and thus make the molecular weight of the organic polymer and the ratio of the inorganic and organic component relatively fixed, this route provide a convenient adjusting of the ratio of the inorganic and organic component and thus the membranes with flexible compositions and properties can be achieved. As shown in Table 10, as the γ-MPS component increased in the copolymer, Td (weight loss of TGA thermogram at 5 wt%) value of the hybrid materials and water contact angle of the hybrid membrane generally increased, while the anion exchange capacity (IEC) and water uptake (W_R) decreased. The charge transition point of the hybrid membranes deduced from their streaming potential (SP) behavior decreased from pH>12 to pH=7-8 as the γ-MPS component increased.
5) The above anion exchange membrane were prepared by sol-gel in aqueous solution. Some times, sol-gel process for anion exchange membrane can also be prepared in non-aqueous solution. As an example, A series of novel hybrid anion exchange membranes were prepared by coupling of 3-glycidoxypropyltrimethoxysilane (GPTMS) and mixed GPTMS/ tetra-n-butyl titanate ($Ti(O\text{-}n\text{-}C_4H_9)_4$) in n-butyl alcohol solution at various temperatures and then followed quaternary amination with trimethyl amine (TMA) [53] (Scheme 14). The anion exchange capacities (IECs) of the membranes decreased with increasing curing temperature within a range of 1.4×10^{-2} to 3.1×10^{-2} meq·cm^{-2} coated for one time and 3.5-4.6$\times10^{-2}$ meq·cm^{-2} coated for two times. The thermal stability of the membrane vary slightly with the content of tetra-n-butyl titanate in the casting solution. As shown in Table 11, with increasing

content of titanium, the starting temperature of weight loss increases; but excessive Ti-content would decrease the stability of Si-O-Ti bonds in the membranes.

In summary, the hybrid anion exchange membranes can be prepared by a variety method. Just like the hybrid materials, in which the most important thing is the precursor, the positively charged precursor is very crucial for hybrid anion exchange membranes. The following steps may be performed by the sol-gel reaction at ambient temperatures, forming metal oxide frameworks by hydrolysis and condensation reactions. The morphologies and properties of the resulting materials are controlled by the reaction conditions and the precursors used [54], consequently they will also influence the properties of anion exchange membranes made from these materials.

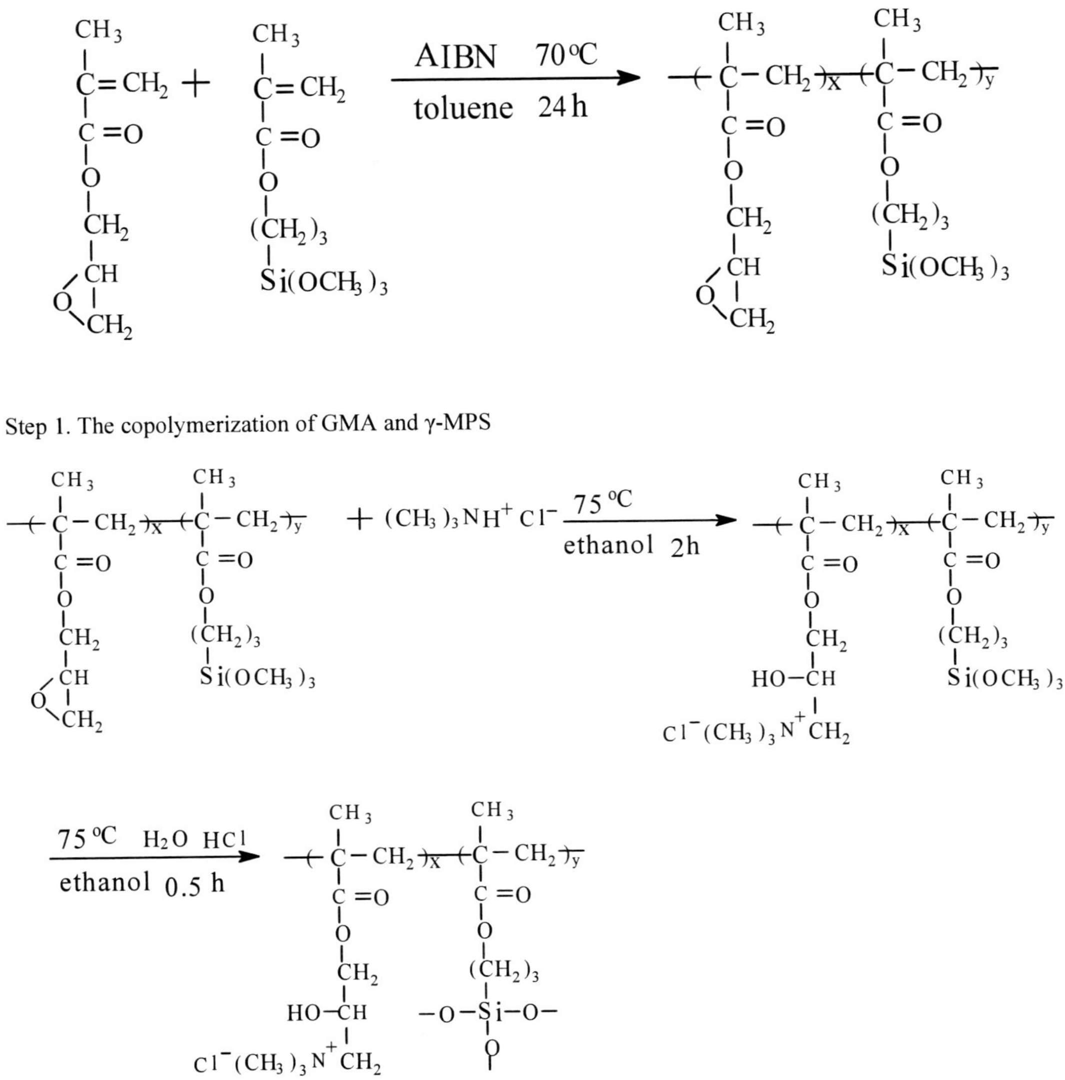

Step 1. The copolymerization of GMA and γ-MPS

Step 2. The ring-opening and sol-gel process of copolymers

Scheme 13. Preparation of anion exchange membrane by copolymerization and sol-gel

Table 10. TGA Analysis Results, Ion Exchange Capacity (IEC), Water Uptake (W_R) and Water Contact Angle of the Hybrid Materials or Membranes in relation with Compositions of the Copolymers

Feed ratio (GMA/ γ-MPS)	Copolymer compositions (GMA/ γ-MPS) [a]	*Td* (°C) [b]	IEC (mmol/g)	W_R(%)	Water contact angle
80/20	83/17	228	1.36	141.4 (±5.2)	83.4°
70/30	71/29	223	1.10	134.87(±1.5)	84.2°
50/50	56/44	241	1.00	82.96 (±2.4)	87.0°
30/70	35/65	251	0.83	79.35(±2.2)	88.2°

[a] Compositions are determined from the ^{1}H NMR spectra of the copolymers
[b] The thermal degradation temperatures (*Td*) are defined as weight loss of TGA thermogram at 5 wt%

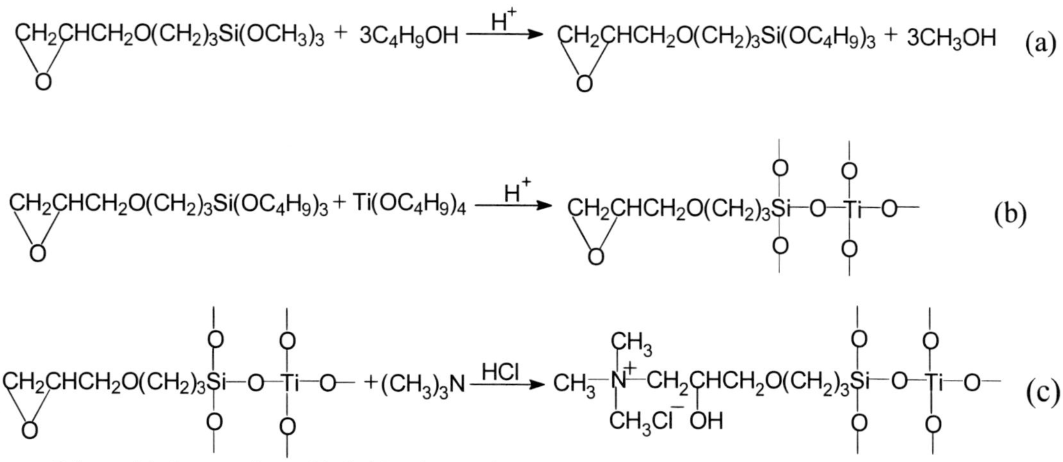

Scheme14. Preparation of hybrid anion exchange membrane by sol-gel in n-butyl alcohol medium. (a) Alcoholysis process; (b) Condensation process;(c) quaternary amination process.

Table 11. The variation of the starting temperature of weight loss with composition of various coating solutions

Coating solutions	GPTMS (mol)	$Ti(O\text{-}n\text{-}C_4H_9)_4$ (mol)	C_4H_9OH (mol)	CH_3COOH (mol)	Ts (°C)
(i)	0.4	0	4	0.01	162
(ii)	0.4	0.09	8	0.01	295
(iii)	0.4	0.15	8	0.01	297
(iv)	0.4	0.30	8	0.01	306

7. Anion Exchange Membranes Preparation by Graft Copolymerization of Vinyl Monomers

Graft copolymerization of vinyl monomers is the most commonly used way to prepare anion exchange membranes instead of chloromethylation of CME or BCME. Such vinyl monomers include pyridine, pyridine-derivatives and vinylbenzylchloride, etc. In most cases, those vinyl monomers were graft-copolymerized onto various polymer films using various grafting technologies such as UV-induce, plasma and irradiation methods. Anion exchange functional groups can be either strongly basic such as tertiary ammonium or weakly basic such as primary, secondary or tertiary amine groups. Anion exchange membranes prepared by grafting of 4-vinylpyridine onto films such as PTFE [55-56], PE [56-57], PP [58-59], ETFE [60], PVDF [61-62], penton [63], polyvinyl chloride (PVC) [64] and poly(hexafluoropropylene-co-tetrafluoroethylene) [65] followed by quaternization with alkyl halide have been reported in literatures. At some cases, crosslinking anion exchange membranes can be obtained by copolymerizing vinyl monomers with other monomers such as divinylbenzene (DVB) to allow the mechanical stability [59,65-67] or to adjust the function of the membranes [68]. For example, two series of anion-exchange membranes have been produced by photo-initiated copolymerization of 4-vinylpyridine and divinylbenzene (DVB) within the pores of polypropylene microfiltration membranes followed by quaternization of the pyridine nitrogen atoms [68]. The first series was comprised of membranes with varying DVB content and all were quaternized with a methyl group. The second series had different quaternizing groups with fixed DVB content. The preparation route was shown in Scheme 15 and the main properties of the membranes as well as their acid recovery performance were shown in Table 12 .It was found that with an increase in DVB content, the first series of membranes showed a decrease in fixed charge concentration, an increase in dialysis coefficients, an increase in selectivity of acid over salt; similarly, with an increase in the size of the *N*-alkyl substituent, the second series of membranes demonstrated an increase in the membrane fixed charge concentration, a decrease in dialysis coefficients and a dramatic increase in selectivity of acid over salts.

8. Halomethylation Process by In-Situ Producing CME or Some Other Reagents

Several other efforts have been made to avoid the use of chloromethyl methyl ether for chloromethylation process. At some cases, maybe the CME is available in the chloromethylation process, but it was produced in-situ by some non-hazardous or less carcinogen materials. For examples, an approach was offered by Amato eta.al. [69], who prepared chloromethylmathylether (CME) free of bis-chloromethylether (BCME), by the reaction of acetylchloride, dimethoxyethane (methylal) in presence of 6 mole % methanol. The ethyl acetate-CME mixture serves as the halomethylating agent. Similarly, iodomethylmethyl ether, free of BCME, was prepared by the reaction of methylal and iodotrimethyl silane [70]. In 1990, Itsuno and coworjers described a very simple and efficient process for the chloromethylation of cross-linked polystyrenes. The method involves the reaction of cross-linked polystyrene with trioxane and chlorotrimethylsilane in the presence of stannic chloride in chloroform. An activated formaldehyde generated from trioxane in the presence of stannic chloride as Lewis acid may react with chlorotrimethylsilane to form a trimethylsilyl ether of the chlorohydrin, having a structure similar to chloromethyl methyl ether, to act as chloromethylating agent for cross-linked poly styrenes. The major advantage of this chloromethylation process is that the reaction was performed in a one-pot manner without contact with the chloromethylating agent generated in the flask. Excess chloromethylating agent can be easily decomposed by a hydrolytic process after the reaction has been completed. The easily decomposable reagent should be much safer than the conventional chloromethylating agents.

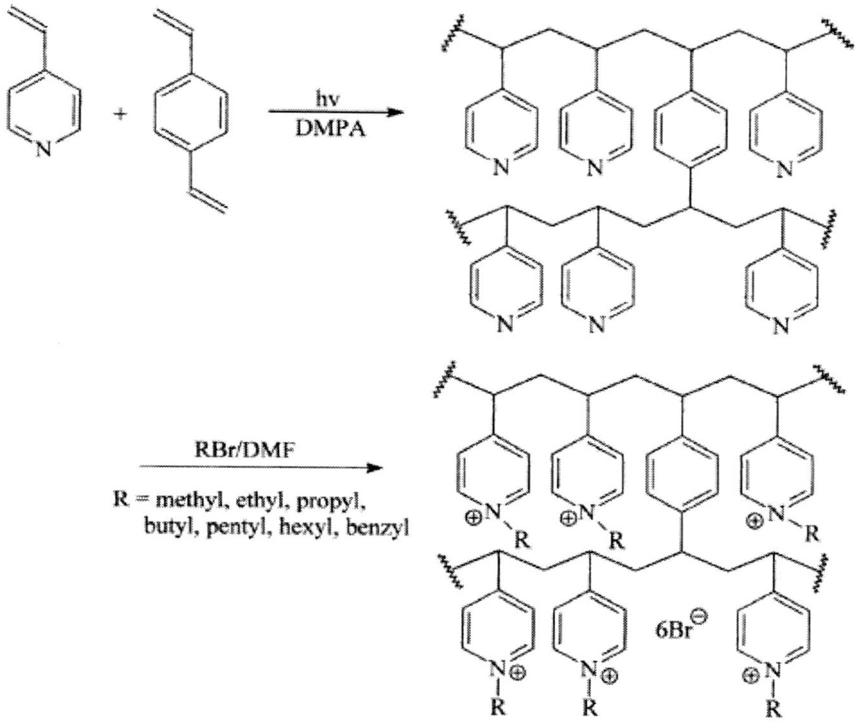

Scheme 15. Preparation of series of anion exchange membranes by photo-initiated copolymerization of 4-vinylpyridine and divinylbenzene (DVB) within the pores of polypropylene microfiltration membranes followed by quaternization of the pyridine nitrogen atoms with *N*-alkyl bromide (RBr), taken from [68]

In addition, some other reagents, such as dimethoxymethane, thionyl chloride and a lewis acid [72], 1-chloro-4-chloro(bromo)- methoxybutane and 1,4-bis[chloro(bromo)-methoxy]butane [[73-74], etc., have been used as new halomethylating agents. Detail information on the new polymeric halomethylating reagents can be seen in a review paper by Warshawsky and Shoef [75].

9. Conclusive Remarks and Perspective

During the last 50 years, anion exchange membranes have evolved from a laboratory tool to industrial products with significant technical and commercial impact. Today anion exchange membranes are receiving considerable attention and are successfully applied for desalination of sea, brackish water and the concentration or separation of food and pharmaceutical products containing ionic species by electrodialysis (ED) , and production of ultra-pure water by electrodialysis-deionzation (EDI) , for cleaning production by Electrodialysis with bipolar membrane (EDBM) [1-5] .

Table 12. The main properties of the membranes with various DVB content or different quaternizing groups as well as their acid recovery performance [68]

Group on nitrogen	DVB (%)	Thickness[a] (μm)	Ion-exchange capacity (meq/g)[b]	Water content[c]	Fixed charge concentration (eq/kg)[d]	U_{HCl}[e]	U_{NaCl}	Selectivity U_{HCl}/U_{NaCl}
Methyl	1.3	119	4.1	64	2.02	67	6.1	36.2
Methyl	4.1	109	4.1	56	2.85	55	3.4	50.7
Methyl	15.4	86	3.6	45	3.78	47	1.3	77.5
Methyl	24.8	85	3.3	41	4.11	34	0.67	111.1
Ethyl	24.8	73	3.2	41	4.59	31	0.4	170.5
n-Propyl	24.8	73	3.0	39	4.91	20	0.18	465.0
n-Butyl	24.8	73	2.9	37	5.18	15	0.088	428.6
n-Pentyl	24.8	73	2.8	36	5.28	9.3	0.02	514.3
n-Hexyl	24.8	76	2.7	26	6.71	6	0.014	9.4
Benzyl	24.8	78	2.6	25	7.25	3.6	0.007	9.1
Hydrogen	0.54	126	4.5	64	2.15	49	5.2	11.1
Hydrogen	1.3	119	4.4	60	2.49	49	5.4	12.8
Hydrogen	2.5	120	4.7	59	2.85	42	3.8	14.0
Hydrogen	4.1	104	4.4	53	3.19	41	3.2	16.2
Hydrogen	6.1	102	4.3	51	3.56	35	2.5	22.5
Hydrogen	8.3	100	4.0	50	3.54	34	2.1	32.1
Hydrogen	15.4	85	3.8	42	4.59	27	1.2	78.1
Hydrogen	24.8	75	3.5	37	5.34	17	0.53	36.2
Hydrogen	44.1	67	3.0	30	6.38	5.7	0.073	50.7

Thickness of membrane in 1M HCl solution after quaternization (±2%).Ion-exchange capacity in meq/g of dry membrane (±3%).Corrected for mass of HCl (±5%).Fixed charge concentration as equivalents of ion-exchange sites per kg of water in the membrane (±5%). U is the dialysis coefficient in mol/hm2M (±5%).

They are also efficient tools for acid recovery by using diffusional dialysis (DD) [26-28,68] or serve along as alkaline solid electrolytes for fuel cell [76-77] and separator for membrane electrolysis [78]. For these important applications , preparation of anion exchange membranes are the most important thing and the routes free of hazardous material such as CME or BCME are very crucial and strongly recommend. Through the above survey, hazardous materials can be avoided in the chloromethylation process by bromination of aromatical polymers with side methyl groups, by Friedel-Crafts reactions, by copolymerization of epoxy group containing monomers or vinyl monomers, by long chain haloalkylethers, by sol-gel and by in-situ production of CME, etc. If the economic and feasibility for scaling-up are considered, the author strongly recommend the bromination route or sol-gel route, because anion exchange membranes produced by bromination is very simple and economically attractive. The intrinsic properties of the obtainable membranes can be quantitatively controlled by the bromination content and position as well as the subsequent crosslingking. Nevertheless, this method is limited in the aromatic polymers only with side alkyl groups(such as $-CH_3$). Sol-gel is a easy route for hybrid material, but it can be used to prepare hybrid anion exchange membranes that can be used in high-temperature occasions such as alkaline solid electrolytes for fuel cell, separator for membrane electrolysis as well as diffusional dialysis at elevated temperature. The other methods reveal more or less flaws such as the expensive monomers or chemicals, the high equipment invest, etc..

It should be also emphasized here, though the above routes decrease the danger in anion exchange membrane preparation, what has been accomplished in the past decade are following the conventional mode, that is, finding the potential applications for a given membrane. To satisfy with a specific requirement, one had to choose a "proper" membrane from the "sea of membranes", optimize the operational condition and then design the process. As strengthened in our recent review on ion exchange membranes [5], in the future, it is necessary to discard this old mode and establish a new one for synthesis and application of anion exchange membranes: the new mode is expected to go the reverse route to the conventional one: designing a specific membrane for a specific purpose. The new mode starts from the needed membrane properties for a specific process, determination of membrane properties, membrane structure and finally design the membrane from molecular level, i.e. such mode goes from molecule to process. Thus, for future preparation of anion exchange membranes, membrane materials-membrane formation-membrane properties –membrane applications should be well mutually correlated in addition to the choice of safe halomethylation routes.

ACKNOWLEDGEMENTS

Financial supports from the Natural Science Foundation of China (No. 20576130, 20376079, 20106015, 29976040), Program for New Century Excellent Talents in University, National Basic Research Program of China (973 program, No. 2003CB615700) and the Special Foundation for Doctoral Discipline of Ministry of Education of China (No. 20030358061) are greatly appreciated. The author would also like to thank his current and former students for their works that are summarized in this paper.

REFERENCES

[1] G. Saracco. Ionic membrane technologies for the recovery of valuable chemicals from waste waters, *Ann. Chim.* Rome. 93 (2003) 817-826.

[2] C.H. Huang, T. W. Xu. "*Cleaner Production by Using Electrodialysis in Organic or Aqueous-Organic Systems*", Edited by Frank Columbus, Nova Science Publishers Inc., Hauppauge, NY 11788, Invited Book Chapter, 2006, In press.

[3] F. N. Xu, C. Innocent and G. Pourcelly. Electrodialysis with ion exchange membranes in organic media. *Sep. Purif. Technol.* 43 (2005) 17-24.

[4] R. Audinos. Ion-exchange membrane processes for clean industrial chemistry. *Chem. Eng. Technol.* 20 (1997) 247-258.

[5] T.W. Xu. Ion exchange membranes: state of development and the perspective, *J. Membrane Sci.* 263 (2005) 1-29.

[6] J. D. Way, R. D. Noble, D. L. Reed, G. M. Ginley, L. A. Jarr. Facilitated transport of CO_2 in ion exchange memberanes, *AIChE J.* 33 (1987) 480-487.

[7] D. Langevin, M. Pinoche, E. Sélégny, M. Métayer, R. Roux. CO_2 facilitated transport through functionalized cation-exchange membranes, *J. Membr. Sci.* 82 (1993) 51-63.

[8] K. Adachi, W. Hu, H. Mastsumoto, K. Ito, A.Tanioka. Permeation of n-butane, 1-butene and 1,3-butadiene through anhydrated Ag^+-doped perfluorocarbon-type ion-exchange, *Polymer* 39 (1998) 2315-2323.

[9] S. Srinivasan, E. A. Ticianelli, C. R. Derouin, A. Rendondo. Advances in solid polymer electrolyte fuel cell technology with low platinum loading electrodes, *J. Power Sources* 22 (1988) 359-375.

[10] H. Wang, G. A. Capuano. Behavior of raipore radiation-grafted polymer membranes in H-2/O-2 fuel cells, *J. Electrochem. Soc.* 145 (1998) 780-784.

[11] Yasuda, N. Yamaga, K. Doi, T. Fujioka, S. Kusanagi. Electrochemical characteristics of the planar electrochemical carbon-monoxide sensor with a perfluorosulfonate ionomer film, *Solid State Ionics* 40-1 (1990) 476-479.

[12] X.F. Li, C.J. Zhao, H. Lu H, Z. Wang, W. Jiang , H. Na. Preparation of a new type of ion-exchange membrane based on sulfonated poly(ether ether ketone ketone)s, *Journal of applied polymer science* 98 (2005) 2481-2486

[13] G. J. Hwang, H. Ohya, T. Nagai. Ion exchange membrane based on block copolymers. Part III: preparation of cation exchange membrane , *Journal of Membrane Science* 156 (1999) 61-65

[14] International Agency for Research on Cancer. Some Aromatic Amines, Hydrazine and Related Substances, Nnitroso Compounds and Miscellaneous Alkylating Agents. Lyon, France: IARC, 1974; *IARC Monogr Eval Carcinog Risks Humans*, Vol. 4: 231-45.

[15] S.Laskin,M. Kusschner, R.T.Drew, V.P.Capiello and N.Nelson. Tumors of the respiratory tract induce by inhalation of bis(chloromethyl)ether, *Arch. Environ, Health* 23(1971)135-136

[16] L.D. Taylor and P.M.C. Laughlin. Preparations and Reations of quaternary salts of bis(chloromethyl)ether, *J.Appl. Polym.Sci.,* 20(1976)2225-2228

[17] P.M. Lang, G. Naumann, US Patent 3989650, 1976.

[18] T. Balakrishnan, W. T. Ford, Particle Size Control in Suspension Copolymerization of Styrene, Chloromethyl Styrene and Divinylbenzene *J. Appl. Polym. Sci.,* 27(1982): 133 - 138.

[19] G.L. Thomas, C. Bohner, M. Ladlow, D.R. Spring.Synthesis and utilization of functionalized polystyrene resins. *Tetrahedron* 61 (2005) 12153-12159.

[20] D. K. Kim, W.J. Chung and Y.S. Lee. Simple bromination of activated arenes by IBX amide resin and tetraethyl-ammonium bromide. SYNLETT (2) (2005): 279-282

[21] E.R. Kenawy. Biologically active polymers. IV. Synthesis and antimicrobial activity of polymers containing 8-hydroxyquinoline moiety. *Journal of Applied Polymer Science* 82 (2001) 1364-1374.

[22] G. Thoma, J.T. Patton, J.L. Magnani, B. Ernst, R. Ohrlein and R.O. Duthaler .Versatile functionalization of polylysine: Synthesis, characterization, and use of neoglycoconjugates *J. American Chem. SOC.* 121 (1999) 5919-5929 .

[23] S. M. Kolhe, A. Kumar. Preparation of strong base anion exchange membrane using Co-60 gamma radiation, *Radiation Physics and Chemistry* 74 (2005) 384-390

[24] T.W. Xu and W.H. Yang. Fundamental study of a new series of anion exchange membranes: Membrane preparation and characterization, *J. Membrane. Sci.*190 (2001) 159-166.

[25] T.W. Xu, F.F. Zha. Fundamental study of a new series of anion exchange membranes: Effect of simultaneous amination-crosslinking processes on membranes' ion exchange capacity and dimensional stability. *J. Membrane. Sci.* 199 (2002) 203-210.

[26] T.W. Xu and W.H. Yang. Sulfuric acid recovery from titanium white (pigment) waste liquor using diffusion dialysis with a new kind of homogenous anion exchange membrane--Static runs, *J. Membrane Sci.* 183 (2001) 193-200.

[27] T.W. Xu and W.H. Yang. Industrial recovery of mixed acid ($HF+HNO_3$) from the titanium spent etching solutions by diffusion dialysis with a new series of anion exchange membranes, *J. Membrane Sci.* 220(2003)89-95.

[28] T.W. Xu and W.H. Yang. Tuning the diffusion dialysis performance by surface crosslinking of PPO anion exchange membranes --simultaneous recovery of sulfuric acid and nickel from electrolysis spent liquor of relatively low acid concentration, *J. Hazardous Material* 109(2004) 157-164.

[29] T.W. Xu , R.Q. Fu,W.H. Yang and Y.H. Xue. Fundamental studies on a novel series of bipolar membranes prepared from poly(2,6-dimethyl-1,4-phenylene oxide)(PPO): (II) Effect of functional group type of anion exchange layers on I-V curves of bipolar membranes. *J. Membrane Science*, 279 (2006) 233-241.

[30] T.W. Xu, Z.M. Liu and W.H. Yang. Fundamental studies of a new series of anion exchange membranes: membranes prepared from poly(2,6-dimethyl-1,4-phenylene oxide)(PPO) and triethylamine, *J. Membrane Science* 249 (2005)183-191

[31] Y. Li, T.W. Xu. and M. Gong. Fundamental studies of a new series of anion exchange membranes: membrane prepared from bromomethylated poly (2,6-dimethyl-1,4-phenylene oxide) (BPPO) and pyridine, *J. Membrane Sci.*, 279 (2006) 151-159.

[32] T. Sata, Y. Tagami. and K. Matsusaki. Transport properties of anion-exchange membranes having a hydrophobic layer on their surface in electrodialysis, *J. Phys. Chem.* B. 102 (1998) 8473-8479

[33] Z. Lin,T. W. Xu and L. Zhang. Irradiation-induced grafting of N-isopropylacrylamide onto the brominated poly(2,6-dimethyl-1,4-phenylene oxide) membranes. *Radiation Chemistry and Physics* 75(2006)532-540.

[34] L. Zhang, T.W. Xu and Z. Lin. Controlled release of ionic drug through the positively charged temperature-responsive membranes, *J. Membrane Sci.*, JMS-05920R2, in press.

[35] H. Xu, X. Z. Hu, Preparation of anion exchangers by reductive amination of acetylated crosslinked polystyrene, *React. Funct. Polym.*, 42 (1999): 235~242.

[36] T. Groth, M. Grotli, W.D. Lubell, L.P. Miranda, M. Meldal. HYDRA: A novel hydroxy and amine functionalised resin synthesised by reductive amination of PEG aldehyde and a polyamine, *J. Chem. Society-Perkin transactions* 1 (2000): 4258-4264

[37] H. Xu and X. Z. Hu.Preparation of Anion Exchange Resins Based on Crossl inked Polystyrene by Friedel - Craf ts Acetylation, *J. Functional Polymers* (In Chinese) 11 (1998) 513~520

[38] L. Wu, T.W. Xu and W.H. Yang. Fundamental studies of a new series of anion exchange membranes: membranes prepared through chloroacetylation of poly(2,6-

dimethyl-1,4-phenylene oxide)(PPO) followed by reductive amination, *J. Membrane Sci.*, JMS-06259, submitted.

[39] D.C. Ge, preparation of IM-2 homogeneous strongly basic anion exchange membrane and its application in desalination of brackish water, *J. Membrane Sci. Tech.* (in Chinese) 9 (1989) 26-30.

[40] K.Sunaga, M. Kim, K. Saito, K. Sugita , T. Sugo. Characteristics of Porous Anion-Exchange Membranes Prepared by Cografting of Glycidyl Methacrylate with Divinylbenzene, *Chem. Mater.* 11(1999) 1986~1989

[41] A.Warshawasky, A. Deshe and N. Gutman. Safe halomethylation of aromatic polymers via BCME-free long chain haloalkylethers, *Br. Polym. J.*16(1984)234-238

[42] Warshawasky, A. Deshe. Halomethyl octyl ethers: convenient halomethylation reagents, *J. Polym.Sci.-Polym Chem.* Ed. 23(1985)1839-1841

[43] A.Warshawasky and O.Kedem. Polysulfone-based interpolymer anion exchange membrane. *J. Membrane Sci.* 53(1990)37-44

[44] A.Warshawasky, N.Kahana, A.Deshe, H.E. Gottlieb and R. ArAD-Yellin. Halomethylated polysulfone:Reactive intermediates to Neutral and ionic film-forming polymers. *J. Polymer Sci., Part A .Poly Chem.* 28(1990)2885-2905

[45] L.L. Hench, J.K. West. The sol-gel process. *Chem Rev* 90(1990):33-72.

[46] M. Kogure, H. Ohya, R. Paterson, M. Hosaka, J. Kim, S. McFadzean, Properties of new inorganic membranes prepared by metal alkoxide methods. Part II. New inorganic-organic anion-exchange membranes prepared by the modified metal alkoxide methods with silane coupling agents, *J. Membrane Sci.* 126 (1997) 161-169

[47] R.K. Nagarale, G.S. Gohil, V.K. Shahi, R. Rangarajan. Preparation of organic–inorganic composite anion-exchange membranes via aqueous dispersion polymerization and their characterization. *J Colloid Interface Sci* 287(2005)198-206.

[48] C.M. Wu, T.W. Xu, W.H. Yang, Fundamental studies of a new hybrid (inorganic-organic) positively charged membrane: membrane preparation and characterizations, *J. Membrane Sci.* 216 (2003) 269-278

[49] C.M. Wu, T.W. Xu, W.H. Yang, Synthesis and Characterizations of Novel Positively Charged PMA–SiO2 Nanocomposites, *Eur. Polym. J.* 41(2005)1901-1908.

[50] S.L. Zhang T.W. Xu, C.M. Wu, Preparation and characterization of new positively charged hybrid membranes based on poly(2,6-dimethyl-1,4-phenylene oxide) (PPO), *J. Membrane Sci.*269 (2006)142-151.

[51] S.L. Zhang, C.M. Wu, T.W. Xu, W.H. Yang, Synthesis and characterizations of new positively charged organic–inorganic hybrid materials based on Poly(2,6-dimethyl-1,4-phenylene oxide) (PPO)， *J. Solid State Chem.* 178(2005)2292-2300.

[52] Y.H. Wu, C.M. Wu， M. Gong and T.W. Xu. New anion exchange organic-inorganic hybrid materials and membranes from a copolymer of glycidylmethacrylate and γ-methacryloxypropyl trimethoxy silane. *J. Applied Polymer Science*, No. App-2006-03-0844R1, in press

[53] J.S. Liu, T.W. Xu, M. Gong M and Y.X Fu.Fundamental studies of a new hybrid (inorganic-organic) positively charged membrane. II. Membrane preparation via alcoholysis reaction and amination processes of silicone and titanate coupling agents. *J. Membrane Sci.* 264 (2005) 87-96

[54] G. Kickelbick, Concepts for the incorporation of inorganic building blocks into organic polymers on a nanoscale, *Prog. Polym. Sci.* 28 (2003) 83-114

[55] G. Kostov, S. Turmanova, Radiation-initiated graft copolymerization of 4-vinylpyridine onto polyethylene and polytetrafluoroethylene films and anion-exchange membranes therefrom, *J. Appl. Polym. Sci.* 64 (1997) 1469-1475

[56] V. Ya Kabanov, L.P. Sidorova, V.I. Spitsyn, Radiation-induced ionic graft polymerization, *Eur. Polym. J.* 10 (1974) 1153-1158

[57] V.N. Kudryavtsev, A. Shapiro, A.M. Endrikhovska-Bonamur, Radiation graft polymerization of 4-vinylpyridine on poly (vinyl chloride) films, Plastich Massy No.4 (1984) 6-11

[58] Kaur, R. Barsola, B. Misra, Graft copolymerization of acrylonitrile and its binary mixture with 4-vinylpyridine onto isotactic polypropylene powder by preirradiation method, *J. Appl. Polym. Sci.* 56 (1995) 1197-1205

[59] A.M. Mika, R.F. Childs, J.M. Dickson, B.E. McCarry, D.R. Gagnon, A new class of polyelectro;yte-filled microfiltration membranes with environmentally controlled porosity, *J. Membrane Sci.* 108 (1995) 37-56

[60] Elmidaoui, S. Belcadi, Y. Houdus, T. Cohen, C. Gavach. Perfluorinated anion-exchange membranes: preparation and preliminary tests of dialysis, *J. Polym. Sci.* A 30 (1992) 1407-1412

[61] G. Ellinghorst, A. Niemoeller, D. Vierkotten, Radiationinitiated grafting of polymer films—an alternative technique to prepare membranes for various separation problems, *Radiat Phys. Chem.* 22 (1983) 635-642

[62] S. Roualdes, N. Kourda, J. Durand, G. Pourcelly, Plasma-grafted PVDF polymers as anion exchange membranes for the electrotransport of Cr(VI), *Desalination* 146 (2002) 273-278

[63] E. Bittencourt, V. Stannett, J. Williams, H. Hopfenberg, Novel ion-containing reverse osmosis membranes. I. Preparation and selected properties, *J. Appl. Polym. Sci.* 26 (1981) 879-888

[64] Dessouki, E.S. Hegazy, M. El-Dessouky, N. El-Sawy, Anionic membranes obtained by radiation grafting of 4-vinylpyridine onto poly(vinyl chloride), *Radiat Phys. Chem.* 26 (1985) 157-163

[65] H. Herman, R. Slade, J. Varcoe, The radiation-grafting of vinylbenzyl chloride onto poly(hexafluoropropylene-co-tetrafluoroethylene) films with subsequent conversion to alkaline anion-exchange membranes: optimization of the experimental conditions and characterization, *J. Membrane Sci.* 218 (2003) 147-163

[66] T. Sata, Y. Yamane, K. Matsusaki, Preparation and properties of anion exchange membranes having pyridinium or pyridinium derivatives as anion exchange groups, *J. Polym. Sci.* A 36 (1998) 49-58

[67] T. Sata, Anion exchange membrane with viologen moiety as anion exchange groups and generation of photo-induced electrical potential from the membrane, *J. Membrane Sci.* 36 (1996) 121-126

[68] D. M. Stachera and R. F. Childs.Tuning the acid recovery performance of poly(4-vinylpyridine)-filled membranes by the introduction of hydrophobic groups. *Journal of Membrane Science* 187 (2001) 213–225

[69] J.S. Amato, S. Karady, M. Sletzinger and L.M. Weisntock. New Preparation of Chloromethyl Methyl Ether Free of Bis[chloromethyl] Ether. Synthesis, No.12 (1979)970-971

[70] M.E. Jung, M.A. Mazurek and R.M. Lim.A New Efficient Synthesis of Iodomethyl Methyl Ether. *Synthesis,* No.8 (1978)588-589

[71] S. Itsuno, K. Uchikoshi, K. Ito k. Novel method for halomethylation of cross-linked polystyrenes *J. Am. Chem. Soc.* 112 (1990) 8187-8188

[72] M.E. Wright, E.G. Toplikar, S.A. Svejda.Details concerning the chloromethylation of soluble high-molecular-weight polystyrene using dimethoxymethane, thionyl chloride, and a lewis acid - a full analysis.*Macromolecules* 24 (1991): 5879-5880

[73] G.A. Olah, D.A. Beal, S.H. Yu and J.A. Olah, Synthetic Methods and Reactions. X[1]. 1-Chloro-4-chloro(bromo)methoxybutane and ,4-Bis-[chloro(bromo)methoxy]butane: New Convenient Halomethylating Agents. synthesis, No.8 (1974)560-561

[74] G.A. Olah, D.A. Beal and J.A. Olah.Aromatic substitution. XXXVIII. Chloromethylation of benzene and alkylbenzenes with bis(chloromethyl)ether,1,4-bis(chloromethoxy)butane,1-chloro-4-chloromethoxybutane and formaldehyde derivatives, *J. Org. Chem.* 41(1976)1627-1631

[75] Warshawsky, N. Shoef, Novel polymeric halomethylating reagents, *J. Polym. Sci.: Polym. Chem.* 23 (1985)1843-1846

[76] J.R. Varcoe, R.C.T. Slade. Prospects for alkaline anion-exchange membranes in low temperature fuel cells, *Fuel Cells* 5 (2005) 187-200

[77] K. Matsuoka, Y. Iriyama, T. Abe, M. Matsuoka, Z. Ogumi. Alkaline direct alcohol fuel cells using an anion exchange membrane. *J Power Sources* 150(2005)27-31

[78] V. Barmashenko, J. Jorissen. Recovery of chlorine from dilute hydrochloric acid by electrolysis using a chlorine resistant anion exchange membrane, *J. Appl. Electrochem.* 35 (2005) 1311-1319

In: Hazardous Materials in the Soil and Atmosphere
Editor: Robert C. Hudson, pp. 89-115
ISBN 1-60021-250-6
© 2006 Nova Science Publishers, Inc.

Chapter 4

CHEMICAL FRACTIONS AND BIOAVAILABILITY OF HEAVY METALS IN POLLUTED SOILS

***Geeta Tewari*[1], *P. C. Srivastava*[2], *Bali Ram*[3] *and Lalit Tewari*[4]**
Department of Chemistry[1], S.S.J. Campus, Kumaun University, Almora, India
Department of Soil Science[2], Department of Chemistry[3]
G.B. Pant University of Agriculture and Technology, Pantnagar, India
Department of Botany[4], D.S.B.Campus, Kumaun University, Nainital, India

ABSTRACT

In a greenhouse experiment, the effect of different levels of heavy metals applied through inorganic salt sources or enriched organic sludge on the dry matter yield and total uptake of heavy metals by spinach (*Spinacia olerecea* L cv "All green") was studied in three different soils.

Treated soils were also analyzed for pH, 1M $CaCl_2$ – and 0.005 M DTPA- extractable heavy metals and five different operationally defined chemical fractions of heavy metals: F_1 (Soluble + exchangeable, in 0.1M $Sr(NO_3)_2$), F_2 (carbonate bound in 1 M NaOAc of pH 5.0) , F_3 (Oxidizable fraction in 5% NaOCl of pH 8.5), F_4 (Reducible fraction in 0.2M oxalic acid + 0.2M NH_4 oxalate + 0.1M ascorbic acid , pH 3), F_5 (Residual fraction in (HF – $HClO_4$ digest). Except Cd, greater proportion of all heavy metals was found in Fe-Mn oxide bound and residual fractions in all polluted soils. In case of Cd, NaOAc extractable fraction was the highest among all fractions while water soluble and exchangeable fraction was higher in inorganically polluted acidic soil. As compared to inorganic salts, sludge application increased the more labile water soluble and exchangeable Zn in all soils. Inorganic salt application increased water soluble and exchangeable Cd, Cu, Ni and Pb; especially in acidic soil. Water soluble and exchangeable Cd, Cu, Ni and Zn and residual Pb fractions showed significant negative correlation with the soil pH.

Heavy metals applied through inorganic salts drastically reduced the dry yield of spinach, especially in acidic soil. The highest uptake of all heavy metals by spinach was recorded in neutral soil. Total uptake of heavy metals applied through sludge treatment was manifold higher than inorganic salt treatment. The 1M $CaCl_2$ extractable amount of Cd, Cu and Ni showed significant positive correlation with the concentration of these heavy metals in spinach tissue. The 0.005M DTPA extractable Cd, Cu, Ni and Zn showed significant positive correlation with the concentration of these heavy metals in spinach tissue. The content of 0.005M DTPA extractable Ni and Zn also showed significant positive correlation with total uptake of these metals by spinach plant.

INTRODUCTION

The application of inorganic fertilizers and organic (sludge) amendments to agricultural lands as a soil physical conditioner and as a source of plant nutrients, have long been practiced in many countries. Many reports have shown that though these applications improve soil fertility, yet they often impose potential hazards to crops, animals and humans owing to the presence of heavy metals (Dowdy and Larson, 1975; Lund et al., 1976).

Research on the possible adverse effects of heavy metal accumulation in agricultural soils amended with sludge or inorganic salt has mainly focused on the available forms of these heavy metals in relation to their uptake by plants (Silviera and Sommers, 1977). Differences in plant availability and soil extractability from these two sources may be primarily related to differences in their heavy metal load and soil properties. Besides this, the addition of metal salts add only the counter anion, while sludge adds organic matter, other minerals carried in the waste stream and also possibly some Fe and lime which are added during the course of waste treatment processes (Baldwin et al., 1983). Sludge constituents, therefore, could also lower metal ion activity in soil solutions through complexation by soluble ligands or immobilization on colloidal surfaces. Thus, heavy metals may exist in mineral forms, such as precipitates, co-precipitates or occlusions in phyllosilicates and sesquioxide minerals (Rudd et al., 1984).

According to Sposito et al. (1982), the mobility and bioavailability of potentially hazardous heavy metals in agricultural soils are closely related to the chemical transformation of these metals in the soil environment. Several sequential extraction methods have been used to partition metals into chemical fractions defined as soluble, exchangeable, adsorbed, organically bound, oxide bound, occluded and residual and to correlate these fractions with plant concentrations or uptake (Emmerich et al., 1982; Silviera and Sommers, 1977; Sims and Kline, 1991). LeClaire et al., (1984), Sposito et al. (1982), Payne et al., (1988) and Sims and Kline, (1991) suggested that metal extraction procedure could also be an index of bioavailability of heavy metals in sludge amended soils.

The objective of the present study were i) to determine the various chemical fractions of Cd, Cu, Ni, Pb and Zn in soils receiving sludge or inorganic salts, ii) to determine the contents of 0.005M DTPA and 1M $CaCl_2$ extractable heavy metals in treated soils and iii) to correlate metal fractions, 0.005M DTPA and 1M $CaCl_2$ extractable heavy metals with metal concentration in plant tissue and total uptake by spinach.

Materials and Methods

Soils and Sewage Sludge

Surface (0-15 cm) samples of acidic, alkaline and neutral soils were collected from Vivekananda laboratory, Almora, Majhola (about 5 km from Moradabad) and Crop Research Centre, Pantnagar, respectively. Soil pH was determined in 1:2 soil/water suspensions on a Systronics (India) digital pH meter. Soil organic C content was determined by Walkley and Black method and texture was determined by hydrometer method. Calcium carbonate was estimated using acid neutralization method. Cation exchange capacity was measured by the NH_4^+ saturation method given by Chapman (1969). The total metal content of the soils was determined in $HF\text{-}HClO_4$ digests by Atomic Absorption Spectrophotometry as outlined by Page et al. (1982). General properties of soil samples used in this study are given in Table 1.

Table 1. General properties of soil samples used in the study

General properties		Acidic soil	Neutral soil	Alkaline soil
Texture		Sandy clay loam	Clay loam	Clay loam
pH (1:2)		6.13	7.00	9.57
O.C. %		0.94	1.20	0.84
% CaCO3		0.39	0.78	2.22
C.E.C. (c mol kg-1)		11.04	15.75	12.49
Field capacity (%)		29.33	27.23	24.28
Total (mg/kg soil)	Cd	1.62	2.60	3.35
	Cu	34.18	24.00	13.68
	Ni	41.06	36.31	52.04
	Pb	18.93	39.06	23.51
	Zn	134.71	109.61	141.01

Sludge was collected from Karula drain (Moradabad), receiving all the sewage effluents from local metal work industries and municipality. The sludge was enriched with 100 mg Zn, Ni and Pb, 200 mg Cu and 50 mg Cd/kg sludge on dry weight basis, mixed and left for one-month incubation.

Green House Study

Two hundred fifty gram of surface soil was filled in plastic pots (72 pots). The organic (enriched sludge) or inorganic (salts) heavy metal sources were added to each soil so as to provide the following level (Table 2) of added heavy metals in the soil.

Hence, the pot experiment receiving a combination of five metals had 3 soils, 4 heavy metal concentrations and 2 sludge levels. There were three replications in the completely randomized design set up.

After treatments, soil of each pot was thoroughly mixed on a polythene sheet and transferred to plastic pots. After keeping the pots at field capacity moisture content for

30days, a basal dose N, P_2O_5 and K_2O at the rate of 30, 20, 30mg kg^{-1} was applied and mixed well. An aliquot of soil from each pot was saved for the estimation of soil pH, chemical fractions of heavy metals, DTPA and $CaCl_2$ extractable heavy metals.

Table 2. Concentration of added heavy metals through inorganic and enriched organic sources

Metals	Concentration (mg kg^{-1})			
	m_0	m_1	m_2	m_3
Cd	0	5	10	20
Cu	0	20	40	80
Ni	0	10	20	40
Pb	0	10	20	40
Zn	0	10	20	40

For chemical fractions of heavy metals, aliquots (3 g) of moist soil were taken into 50-ml polycarbonate centrifuge tubes and sequentially extracted as per the scheme of Ahnstrom and Parker (1999) to obtain the following fractions: F_1 (Water soluble + exchangeable, in 0.1M $Sr(NO_3)_2$), F_2 (carbonate bound, in 1 M NaOAc of pH 5.0) , F_3 (Oxidizable fraction, in 5% NaOCl of pH 8.5), F_4 (Reducible fraction, in 0.2M oxalic acid + 0.2M NH_4 oxalate + 0.1M ascorbic acid , pH 3), F_5 (Residual fraction in (HF – $HClO_4$ digest). Soil samples were also analyzed for 1M $CaCl_2$- and 0.005 M DTPA -extractable heavy metals following the procedures outlined by Ahnstrom and Parker (1999) and Lindsay and Norvell (1978, respectively). Heavy metal contents in each fraction were estimated by atomic absorption spectrophotometry and the results were expressed on oven dry weight basis.

Six healthy seeds of spinach (*Spinacia olerecea* L. cv "All Green") were sown in each pot. After the emergence, thinning was done to maintain only two plants per pot. The pots were regularly watered to maintain them at field capacity moisture content. After 30 days of growth, plants were harvested, washed sequentially with tap water, 0.1N HCl solution and finally with distilled water. Dry matter yield was recorded after drying the plant samples at 60°C for 48h. Dry plant tissue was finely ground and wet-ashed using HNO_3: H_2SO_4: $HClO_4$ (10:1:4v/v). The concentration of heavy metals in the acid extracts was determined by atomic absorption spectrophotometry.

Statistical Analysis

Sewage sludge and inorganic salt treatment effects on dry yield, 1 M $CaCl_2$- and DTPA –extractable contents and chemical fractions of heavy metals in soil, concentrations and total uptake of heavy metals in lettuce were analyzed based upon completely randomized factorial design with three replications using variance analysis (ANOVA). A probability level of p=0.05 was selected to establish statistical significance. Correlation and stepwise multiple regression analyses were also performed with the data to correlate concentration of metals in different chemical fractions, plant concentration and total uptake.

RESULTS AND DISCUSSION

The enriched sludge had 5.2 pH and 48 gm organic C. After enrichment, the final total content of metals in the sludge was found to be 257.6 mg Ni, 2225.8 mg Zn, 1517.4 mg Cu, 53.8 mg Cd kg^{-1} sludge.

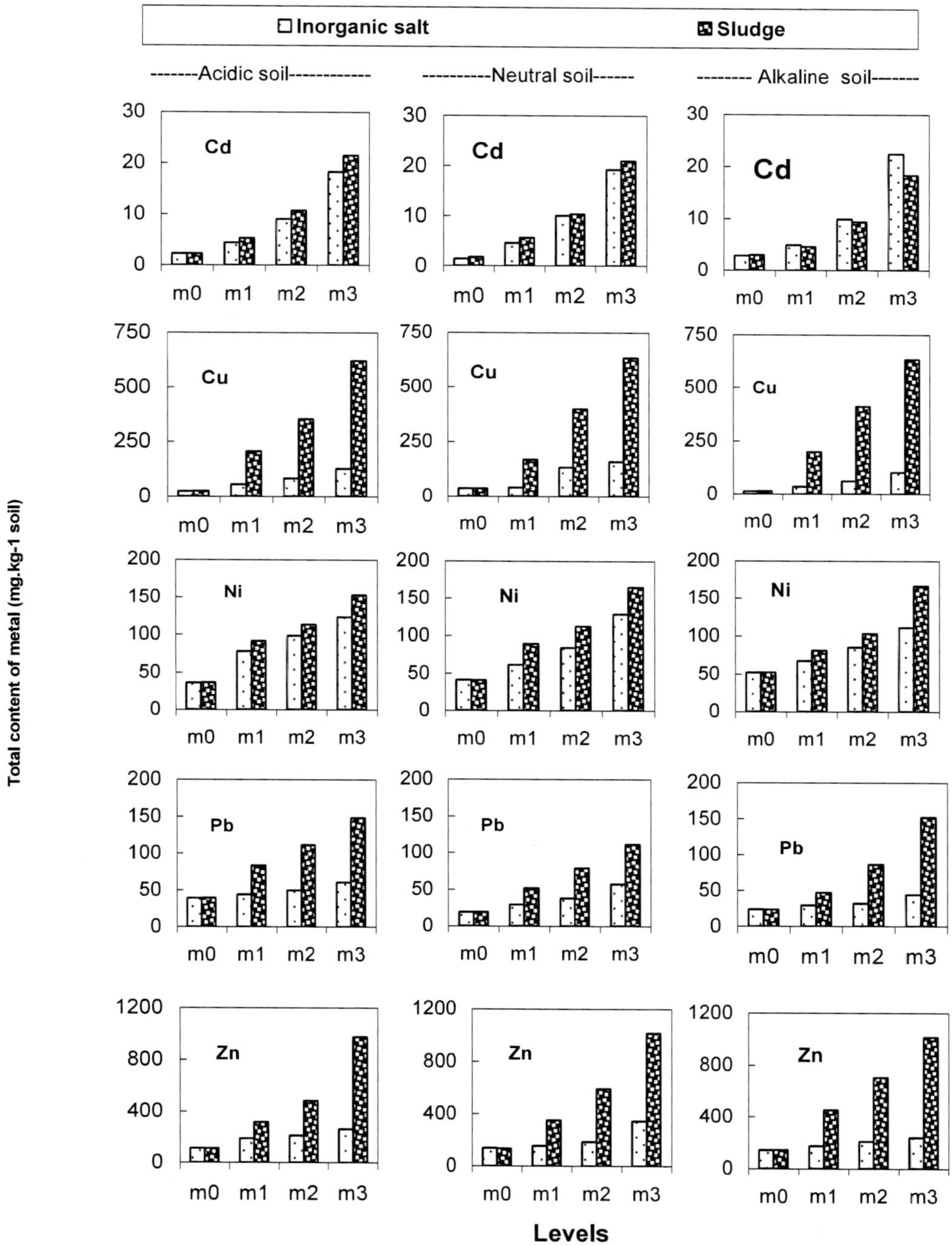

Fig. 1. Total content of heavy metals in soils under different levels of inorganic salts and sludge additions.

Total Content of Heavy Metals in Soils

Total contents of heavy metals in soils treated with inorganic salts and enriched sludge are depicted in Fig. 1. Total Cd contents in the soils spiked through inorganic salt or enriched sludge were relatively very close as Cd was originally not present in high concentration in the sludge. However, the total contents of rest other heavy metals, especially that of Cu, Zn and Pb were much higher with the addition of enriched sludge than inorganic salts due to high content of these heavy metals in the sludge.

Chemical Fractionation of Heavy Metals in Soils

Heavy metals present in sludge or other organic wastes and inorganic fertilizers may increase their total levels of in agricultural soils. Their use in agriculture could alter the intensities of various chemical pools of metals in the soil and consequently affect the bioavailability of these metals in the environment. The contents of different chemical fractions of heavy metals in three soils treated with inorganic salts or sludge are shown in Fig. 2 to 6.

In order to evaluate the relative distribution of different chemical fractions of heavy metals in soils receiving inorganic metal salts or sludge, the percent distribution of various chemical fractions of the metals was also calculated and depicted in Fig. 7. It was observed that in control (m_0 level), most of the Cd was present in residual fraction (62-88%), which is supposed to be a part of crystal lattice of soil minerals and has poor mobility. This fraction is believed to be least important as far as plant availability in a short period of time under normal conditions is concerned (Tessier et al., 1979). In soils treated with Cd, soluble + exchangeable-, NaOAc extractable- and organically bound- fractions together accounted at least 66% of total metal content. This suggests that Cd in treated soils was present in potentially bio-available form as the metal present in these pools was likely to be available for plant uptake (Xian, 1989). In soils polluted by Cd though acidic sludge or inorganic salts, Cd was distributed among NaOAc extractable-, organically bound- and residual fractions. Similar distribution patterns have been observed by several other workers (Sposito et al., 1983; Silviera and Sommers, 1977) who also noted that in sludge amended soils, the greater percentage of Cd was present in NaOH- (organically bound), EDTA- (precipitated) and HNO_3 – extractable fractions. The mean percentage of water soluble + exchangeable fraction was the highest in acidic soil which could be attributed to lower soil pH of amended soils. Lower percentage of water soluble + exchangeable- and NaOAc extractable- fractions in organically polluted soil as compared to inorganically polluted ones might be attributed to complexation of added Cd as organo-metallic compounds.

The most abundant pool for Cu in treated soils was non-residual pool (42.4% to 64.4%) except in neutral soil especially, with addition of inorganic source of metal (Fig. 7). Among non-residual fractions, the Fe-Mn oxide bound pool was the most dominant (37.0% to 47.2%) possibly owing to co-precipitation of added Cu with iron and Mn oxides present in soil. In soils, Fe and Mn oxides are known to exist as discrete particles or as coating on particles and have also been reported to be excellent scavengers for heavy metals both in natural and polluted soil (Kuo et al. 1983 Xian, 1989), however, Fe-Mn oxides are unstable under submerged and low Eh conditions (Tesssier et al., 1979). Only a small percentage of Cu was found to exist as soluble + exchangeable fraction (0.23% to 1.14%). On an average the percent distribution of different Cu fractions followed the order:

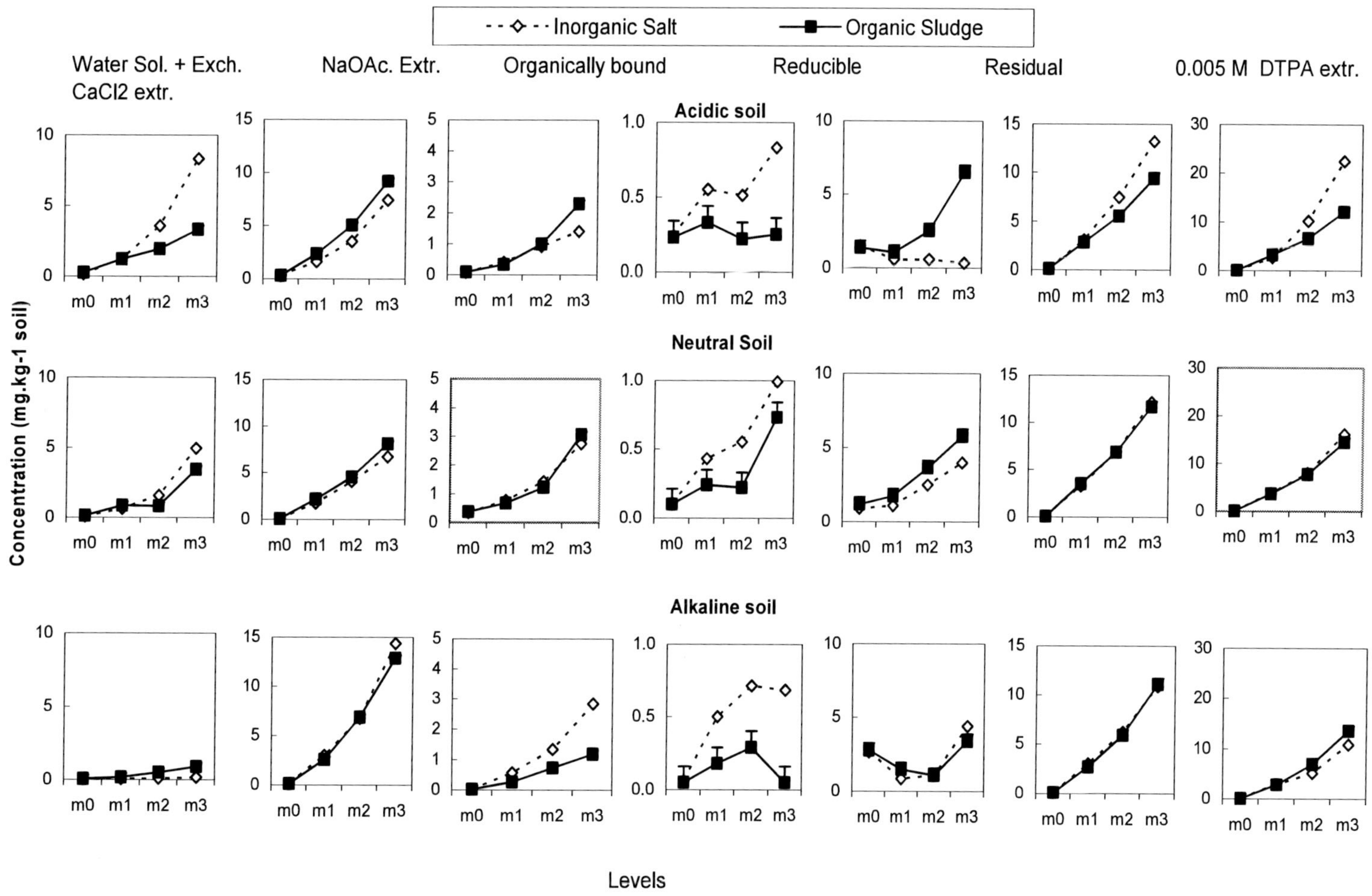

Fig. 2. Effect of different levels of Cd added through inorganic salts and organic sludge on chemical fractions of Cd in soils.

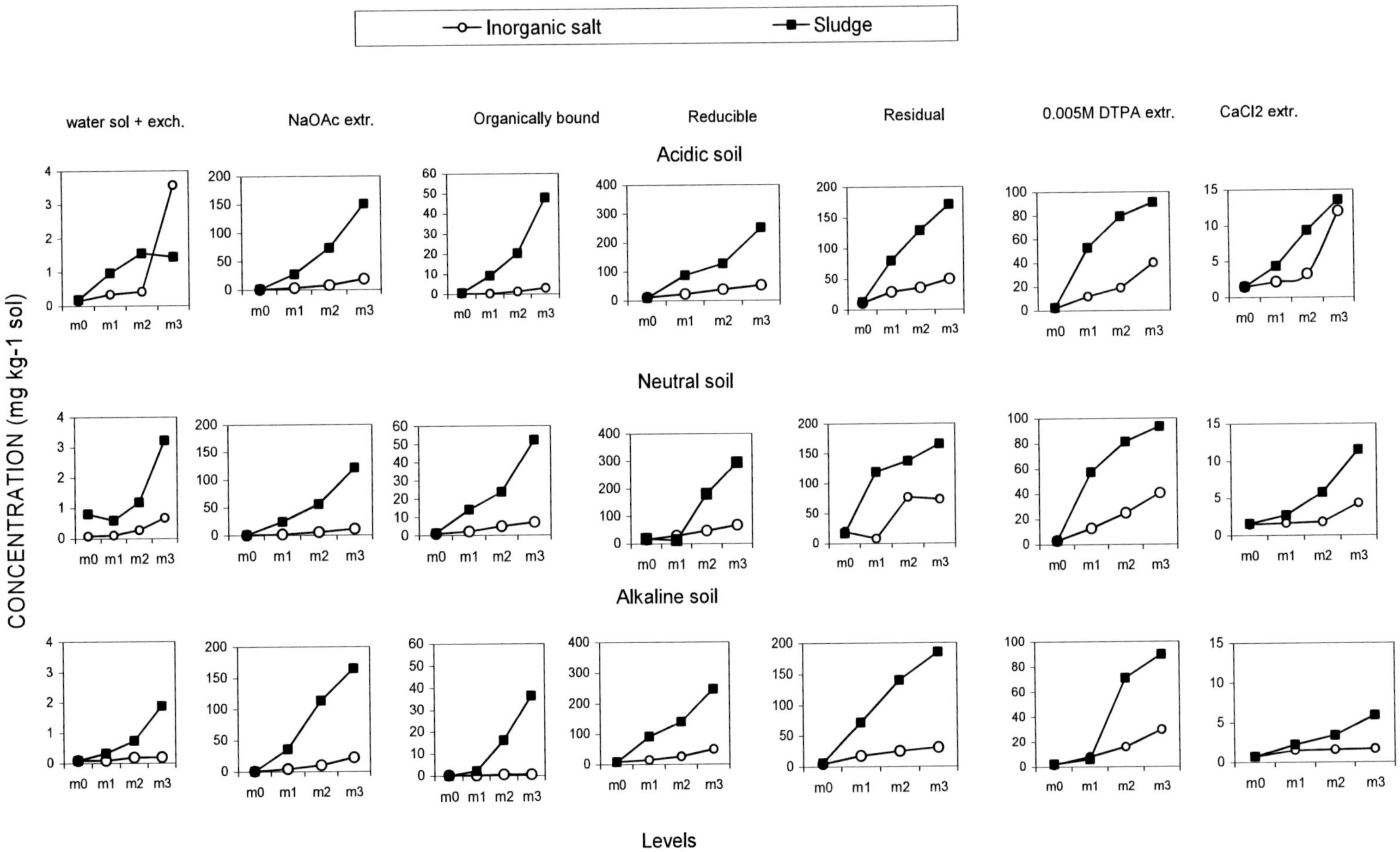

Fig. 3. Effect of different levels of Cu added through inorganic salts and organic sludge on chemical fractions of Cu in soils.

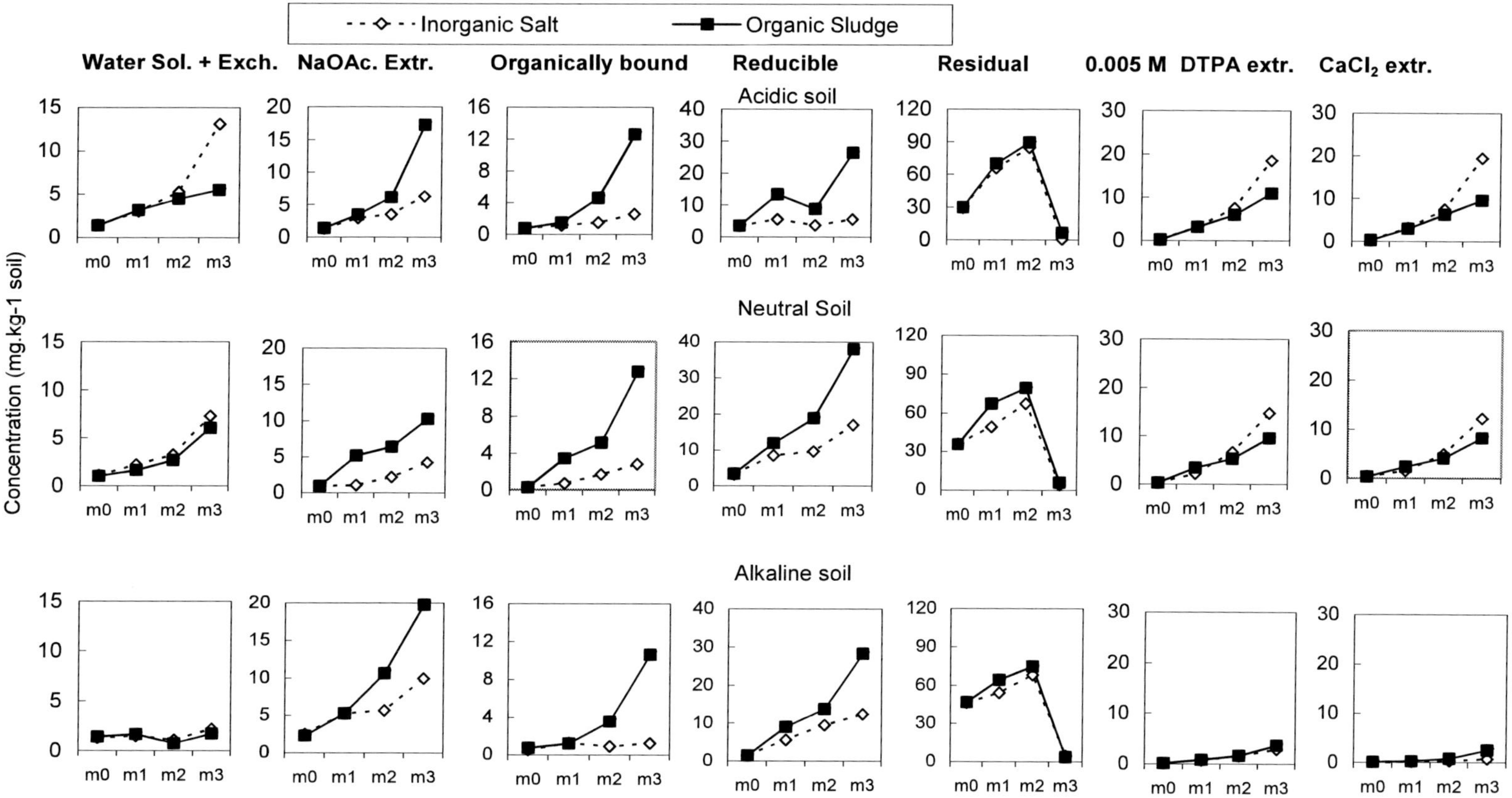

Fig. 4. Effect of different levels of Ni added through inorganic salts and organic sludge on chemical fractions of Ni in soils.

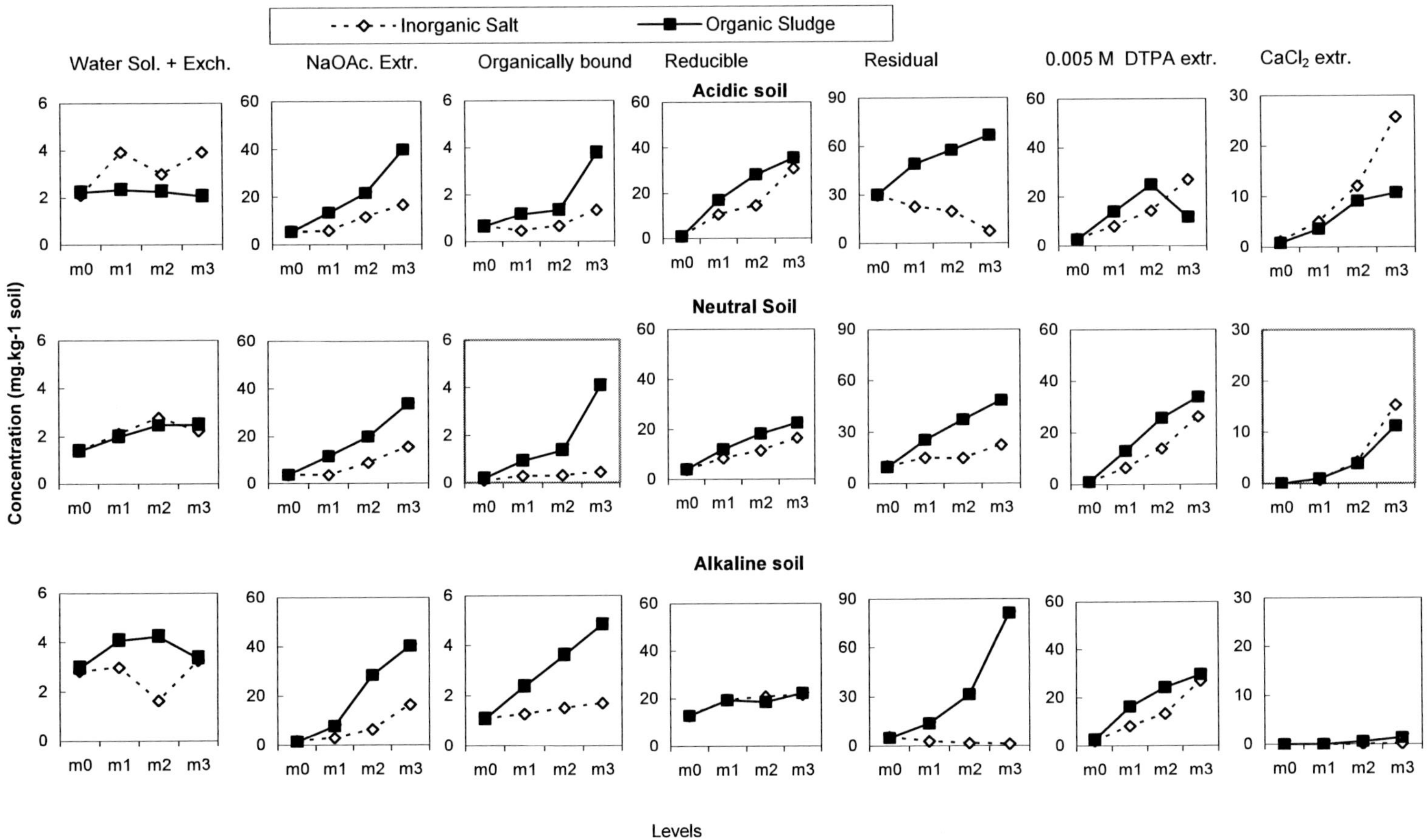

Fig. 5. Effect of different levels of Pb added through inorganic salts and organic sludge on chemical fractions of Pb in soils.

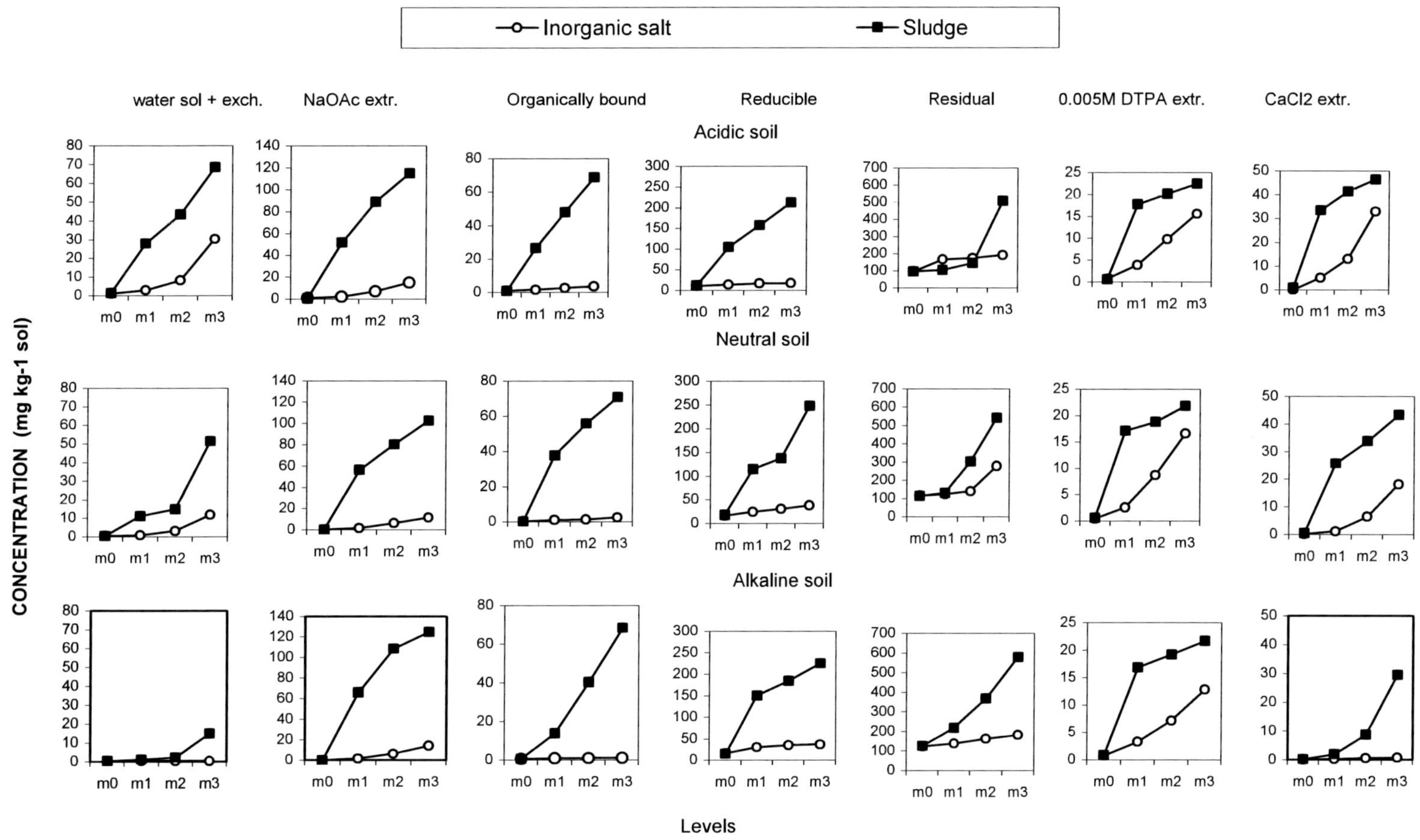

Fig. 6. Effect of different levels of Zn added through inorganic salts and organic sludge on chemical fractions of Zn in soils.

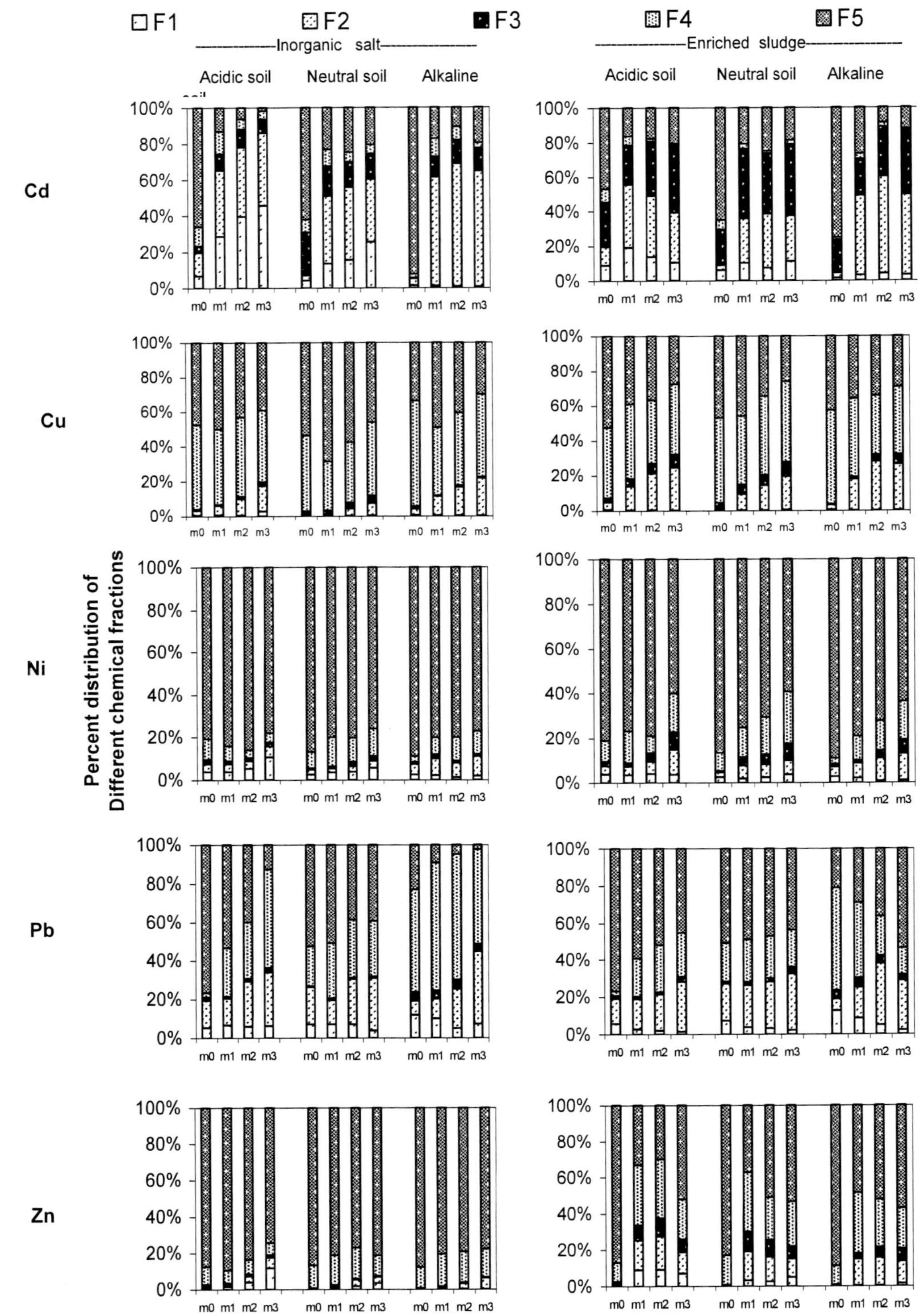

Fig. 7. Effect of inorganic and organic (sludge) sources on per cent distribution of different chemical pools of heavy metals in soils.

Fe-Mn oxide bound > residual > NaOAc extractable > organically bound > soluble + exchangeable

Application of acidic sludge increased NaOAc extractable and organically bound fractions of soil Cu. This might be due to the reason that organic matter and $CaCO_3$ present in sewage sludge served both as a source of Cu and also as major adsorbent for this metal in soils. The results of this study were found to be consistent with the results of Sims and Kline (1991) and Morera et al. (2001) who also reported that with sewage sludge application increased organically bound fraction of Cu in soil. In soils polluted by inorganic Cu salts, only a small change was observed for exchangeable Cu. These results corroborated with the findings of Chlopecka (1993). In treated soils, percentage of NaOAc extractable Cu fraction increased at the expense of residual and Fe-Mn oxide bound fraction.

In the case of Ni, the most abundant chemical pool in untreated soil was residual (80-89% of total Ni). These results were consistent with the observations of Gupta and Chen (1975) and Tessier et al. (1979) who suggested that most of Ni in soils and sediments was detritus in nature. Among non-residual pools in non-polluted soils, Fe-Mn oxide bound Ni fraction was the most abundant pool (2-13% of total Ni). Nickel distribution changed when soils were amended with sludge and inorganic salts. In sludge amended soils, there was a significant increase in the NaOAc extractable-, organically- and Fe-Mn oxides- bound Ni fractions but the percentage fraction of residual Ni decreased. Taylor et al. (1995) reported similar results for Ni in sludge-amended soil. Sposito et al. (1983) also observed an increase in the organically bound Ni fraction with an associated decrease in the residual fraction at increasing rates of sludge application.

In untreated soils except alkaline soil, the highest percentage of Pb was found in residual fraction (52-76% of total) due to higher association of Pb with this fraction and little redistribution of this metal from native residual Pb fraction in soil. In soils polluted by sludge, the NaOAc extractable-, organic bound- and Fe-Mn oxides bound- Pb fractions increased while only a slight decrease occurred in the residual- and water soluble + exchangeable- Pb fractions. This indicated a higher affinity of Pb for carbonates and Fe-Mn oxides. Similar results have been reported by Sposito *et al.* (1983) and Taylor *et al.* (1995).

With addition of Pb through inorganic salts, only small change occurred in the percentage of water soluble + exchangeable- and organically bound- Pb fractions in all soils except in case of alkaline soil. The percentage of NaOAc extractable- and Fe-Mn oxides bound- Pb fractions in soils treated with inorganic salt also increased significantly while the relative proportion of the metal in residual fraction was very small in the treated soils. This might be due to precipitation or/and co-precipitation of Pb into carbonates and Fe-Mn oxides present in soil or newly formed iron-oxides. A similar pattern of transformation was observed by Chlopecka (1993) for Pb in polluted soil.

In these soils, Zn was mostly concentrated in the residual fraction, though it was also present in significant amounts in other fractions as well. The percent distribution of Pb in the residual fraction ranged from 50.5% to 83.7%. The greater percent distribution of Zn in the residual fraction indicated the higher tendency of this metal to become unavailable in soils. Similar results were reported for a soil from tailing dumps (Jordao and Nickless, 1989), near-shore sediments (Gupta and Chen, 1975) and contaminated soils (Bell et al., 1991; Ma and Rao, 1997). The % amount of Zn present in other chemical fractions ranged from 2.3% to 49.5%. Among non-residual chemical fractions of Zn, the Fe-Mn oxide bound fraction (7.9%

to 24.4%) was the next most dominant fraction in all soils. Several other workers have also reported significant amount of soil Zn to be associated with Fe-Mn oxides (Kuo et al., 1983; Ramos et al., 1994; Ma and Rao, 1997). Various chemical fractions of Zn in soils polluted through sludge and inorganic salts could be arranged in the following order:

Organic source: Residual > Fe-Mn oxide > NaOAc extractable > Organically bound > Soluble + exchangeable fraction

Inorganic source: Residual > Fe-Mn oxide > NaOAc extractable >Soluble + exchangeable > Organically bound fraction

Although increasing level of heavy metals applied through inorganic salts and acidic sludge increased the amount of Zn in all chemical fractions, the percentage of metal in F1, F2, F3 and F4 fractions increased significantly while in F5 (residual) a significant decrease was observed. The decrease in % of residual fraction might be due to preferential redistribution of soil Zn into other mobile fractions when soils were amended with acidic sludge. Release of Zn into bioavailable (water soluble + exchangeable- and NaOAC extractable-) forms, could be controlled either by complexation or by precipitation. The increase of Zn in Fe-Mn oxide bound fractions could be due to co-precipitation of released Zn with newly formed oxides. Iron and manganese oxides are known to be an important sink for the added Zn. The greater proportion of Zn bound to Fe and Mn oxides in the sludge-amended soils as compared to inorganic salt treated soils suggested that a higher proportion of Fe and Mn in sludge amended soils existed in a more reactive para-crystalline hydrous oxide state. Similar trends were also noted by LeClaire et al., (1984) and Sims and Kline (1991) who also reported that percentage of organic and precipitated fractions of Zn were increased following amendment of soil with composted sludge. Through inorganic salts, the percentage of F1, F2 and F3 fractions increased while F4 and F5 fractions decreased (except increase in F4 fraction in alkaline soil). The increase in bioavailable fraction was the highest in acidic soil probably due to low pH. These results suggested that Zn in inorganic salt polluted soil could be easily mobilized to exchangeable, NaOAc extractable and organically bound fraction at the expense of from residual- and Fe-Mn oxide bound- Zn fractions.

The Relationship Between Different Chemical Fractions of Heavy Metals and Total Metal Content and Soil pH in Polluted Soils

Simple correlation coefficients (r) between different chemical fractions of heavy metals and total content of these heavy metals in polluted soils and soil pH are given in Table 3.

All chemical fractions of heavy metals showed significant and positive correlation with the total content of respective metals in polluted soils except for soluble + exchangeable fraction of Pb and iron and manganese oxide bound fraction of Cd.

As regards the relationship between soil pH and different chemical fractions of heavy metals, water soluble and exchangeable fraction of Zn, Ni, Cu and Cd showed significant negative correlation with soil pH, however, such relationship was not significant in case of Pb. This relationship is in agreement with the known fact that soil pH is the most important single factor controlling the metal mobility (Adriano, 1986). Higher pH increases the co-precipitation, occlusion or strong adsorption of these heavy metals. The similar results were reported by Soon and Bates (1982) and Dudka et al. (1996). The residual fraction of Pb showed a significant negative correlation with soil pH.

Table 3. Simple correlation coefficient between different chemical fractions of heavy metals and total heavy metal content and soil pH in polluted soil

Heavy metals	Extractable	Chemical fractions of heavy metals				
		F_1	F_2	F_3	F_4	F_5
Cd	Total content in soil	0.661**	0.862**	0.855**	0.334	0.865**
	pH	-0.696**	-0.038	-0.337	-0.348	-0.181
Cu	Total content in soil	0.869**	0.960**	0.961**	0.964**	0.965**
	pH	-0.447*	-0.169	-0.358	-0.311	-0.365
Ni	Total content in soil	0.656**	0.851**	0.866**	0.926**	0.968**
	pH	-0.669**	0.019	-0.316	-0.395	-0.304
Pb	Total content in soil	0.163	0.940**	0.811**	0.727**	0.934**
	pH	0.280	-0.270	0.095	0.046	-0.520*
Zn	Total content in soil	0.727**	0.944**	0.939**	0.960**	0.952**
	pH	-0.455*	-0.194	-0.289	-0.198	-0.189

Interrelationship Among Different Chemical Fractions of Heavy Metals in Polluted Soils

The matrixes of simple correlation coefficients (r) among different chemical fractions of heavy metals in polluted soils are presented in Table 4.

It is evident from the 'r' values that all the chemical fractions of Zn and Cu were significantly and positively correlated among themselves indicating thereby the existence of dynamic equilibrium among them. In case of Cd, a similar significant positive relationship existed among different chemical fractions except between water-soluble + exchangeable and NaOAc extractable fraction; and iron and manganese oxides bound and residual fractions. In case of Pb, except for water soluble + exchangeable fraction which failed to show a significant correlation with any other chemical fraction of Pb, rest other chemical fractions were significantly and positively correlated among themselves.

Greenhouse Experiment

Application of different levels of heavy metals to soil through sludge and inorganic salts influenced the growth of spinach plants (Plate 1). No plant growth occurred in treatment that received the highest level of heavy metals through inorganic salts in the acidic soil.

Table 4. Matrix of simple correlation coefficient (r) among different chemical fractions of heavy metals in polluted soils

Heavy metals	Chemical fraction	Chemical fractions of heavy metals				
		F_1	**F_2**	**F_3**	**F_4**	**F_5**
Cd	F_1	1.000	0.267	0.557**	0.610**	0.653**
	F_2		1.000	0.831**	0.781**	0.768**
	F_3			1.000	0.919**	0.723**
	F_4				1.000	0.819**
	F_5					1.000
Cu	F_1	1.000	0.789**	0.910**	0.894**	0.814**
	F_2		1.000	0.906**	0.932**	0.904**
	F_3			1.000	0.950**	0.896**
	F_4				1.000	0.894**
	F_5					1.000
Ni	F_1	1.000	0.355	0.715**	0.542**	0.654**
	F_2		1.000	0.787**	0.417	0.613**
	F_3			1.000	0.697**	0.786**
	F_4				1.000	0.217
	F_5					1.000
Pb	F_1	1.000	0.170	0.408	0.250	0.035
	F_2		1.000	0.844**	0.763**	0.792**
	F_3			1.000	0.689**	0.638**
	F_4				1.000	0.473*
	F_5					1.000
Zn	F_1	1.000	0.693**	0.783**	0.722**	0.601**
	F_2		1.000	0.954**	0.983**	0.812**
	F_3			1.000	0.949**	0.815**
	F_4				1.000	0.839**
	F_5					1.000

** significant at p=0.01; * significant at p=0.05

Dry Yield

The data on the effect of heavy metals added through inorganic salts or acidic sludge on dry weight of spinach grown in different soils are shown in Table 5.

It was evident from the data that the highest dry weight of spinach was recorded in neutral soil followed by alkaline and acidic soil. Higher dry weight was recorded for the soils treated with sludge as compared to inorganic salts. The higher growth in sludge amended soil may have been caused by improved nutrition and water holding capacity from organic matter added as sludge.

The interaction effect of soil x source x heavy metal concentration was found to affect the dry weight of spinach significantly. In acidic soil, heavy metals added through inorganic salts decreased the dry weight of spinach significantly in comparison to the control and the dry weight was drastically reduced at m_2 level and above however, when sludge was used as the source of heavy metals, an increased dry weight of spinach was observed up to m_2 level in comparison to control and beyond this level a significant decrease in dry weight was registered. In alkaline soil, heavy metals added through inorganic salts showed significant successive decrease in the dry weight of spinach in comparison to the control. When sludge was the source of heavy metals, a significant decreased dry weight of spinach was observed up to m_2 level and beyond this level; dry weight of spinach was increased non-significantly as compared to the control (m_0). In neutral soil, heavy metals added through inorganic salts increased the dry weight of spinach significantly in comparison to the control (m_0). When sludge was the source of heavy metals, the increase in dry weight was significant up to m_1 level and non-significant increase was observed at m_2 level as compared to the control (m_0). Beyond m_2 level, significant decrease was registered in the dry weight of spinach as compared to the control (m_0).

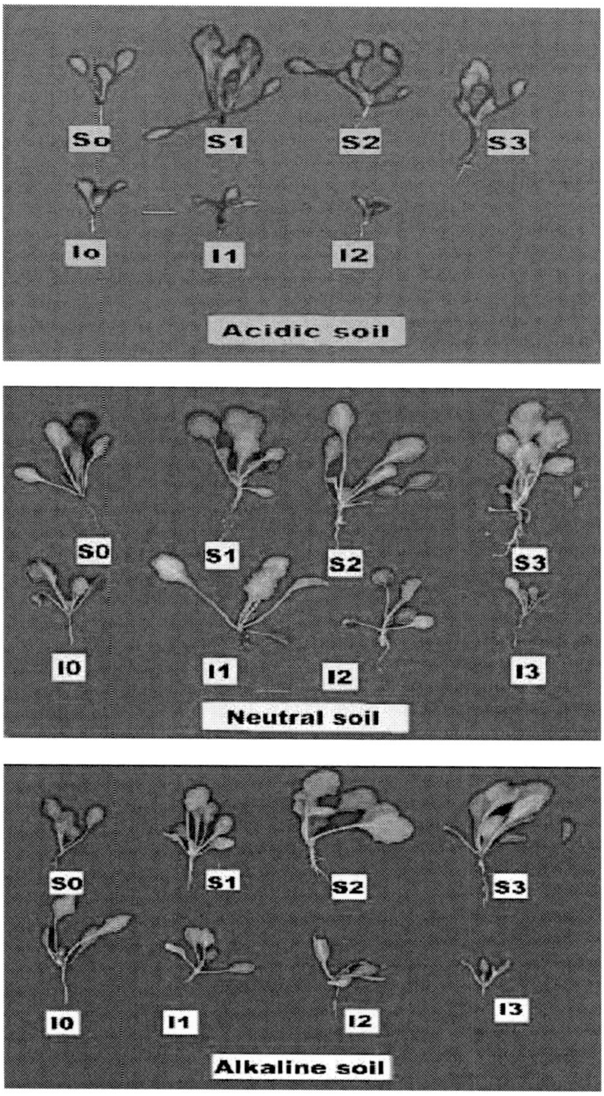

Plate 1. Effect of different levels of heavy metals added through sludge (S) and inorganic salts (I) on the growth of Spinach.

Table 5. Effect of heavy metals added through inorganic salts or enriched acidic industrial sludge on dry yield of Spinach (mg/plant) grown in different soils

Conc	Neutral Soil			Alkaline soil			Acidic Soil			Mean		
	I	O	Av	I	O	Av	I	O	Av	I	O	Av.
m_0	0.13	0.18	0.15	0.13	0.12	0.13	0.09	0.09	0.09	0.12	0.13	0.12
m_1	0.15	0.34	0.25	0.08	0.09	0.09	0.05	0.12	0.08	0.10	0.18	0.14
m_2	0.14	0.35	0.24	0.06	0.10	0.08	0.01	0.12	0.07	0.07	0.19	0.13
m_3	0.06	0.13	0.10	0.03	0.13	0.08	0.01	0.03	0.02	0.04	0.10	0.07
Av	0.12	0.25	0.19	0.08	0.11	0.09	0.04	0.09	0.07	0.08	0.15	
Effects	Soil (Sl)	Sludge (Sd)	Conc.	Sl x Sd	Sd x C	Sl x C	Sl x Sd x C					
SEm	0.002	0.002	0.002	0.003	0.004	0.004	0.006					
CD (p=0.05)	0.006	0.005	0.007	0.009	0.012	0.009	0.017					

I=Inorganic salts and O=organic enriched, acidic sludge

Among soils, irrespective of source of heavy metal addition the dry matter yield of spinach was the highest in neutral soil. This might be due to presence of high organic matter in soil itself, which could contribute higher nutrient supply.

Decrease in dry yield of spinach in alkaline soil polluted with inorganic as well as organic sources might be due to some micronutrient deficiencies at higher pH values (Sims and Kline, 1991).

The low pH and low CEC of acidic soil must have favoured the conditions for the higher accumulation of heavy metals in this soil, which in turn decreased the dry yield. The high pH and CEC of alkaline soil might be responsible for lowering the toxicity of heavy metals in this soil. The present results corroborated with earlier reports of John et al. (1972), Haghiri (1974), Mahler et al. (1978), Street et al. (1978) and Ramachandran and D'Souza (1999), where reduced availability of Cd to plant shoots was demonstrated in soils having high pH and high CEC. Valdares et al., (1983) also found a negative correlation between dry yield of Swiss chard and the total metal added through sludge to a non-calcareous soil.

Concentration of Heavy Metals in Spinach and Their Total Uptake

The major environmental concern of heavy metals contamination through pollutants is their uptake by plants (especially food crops) leading to their entry into food chain and bio-magnification and ground water contamination through possible leaching. Metal uptake depends on the plant species and on the chemistry of the metals in the soil. Elevated metal content in crop plants irrigated with sewage/ industrial effluent or application of metal containing sewage sludge or other inorganic bye-products has been reported by Chang et al. (1983); Narwal et al. (1983); Valdres et al. (1983); Pietz et al. (1983); Srikanth and Reddy, (1991) and Cunningham et al. (1975).

The effect of heavy metals added through inorganic salts or sludge on their concentration in spinach plant are depicted in Fig. 8. Total uptake values of heavy metals by spinach grown in different soils are shown in Table 6-10.

Table 6. Effect of heavy metals added through inorganic salts or enriched acidic sludge on uptake of Cd by spinach tissue (µg/plant)

Cu	Neutral soil			Alkaline soil			Acidic soil			Mean		
	I	O	Av	I	O	Av	I	O	Av	I	O	Av.
m_0	0.35	0.48	0.42	0.89	0.83	0.86	1.82	2.01	1.91	1.02	1.11	1.06
m_1	2.31	11.59	6.95	1.69	2.03	1.86	1.55	4.14	2.84	1.85	5.92	3.88
m_2	2.93	19.84	11.38	0.92	2.95	1.93	0.01	5.85	2.93	1.28	9.55	5.42
m_3	2.32	3.80	3.06	0.15	3.62	1.88	0.01	1.52	0.77	0.83	2.98	1.90
Av	1.98	8.93	5.45	0.91	2.36	1.63	0.85	3.38	2.11	1.24	4.89	
Effects	Soil(Sl)	Sludge (Sd)	Conc.	Sl x Sd	Sd x C	Sl x C	Sl x Sd x C					
SEm	0.16	0.13	0.18	0.23	0.32	0.26	0.45					
CD (p=0.05)	0.46	0.37	0.51	0.65	0.91	0.74	1.28					

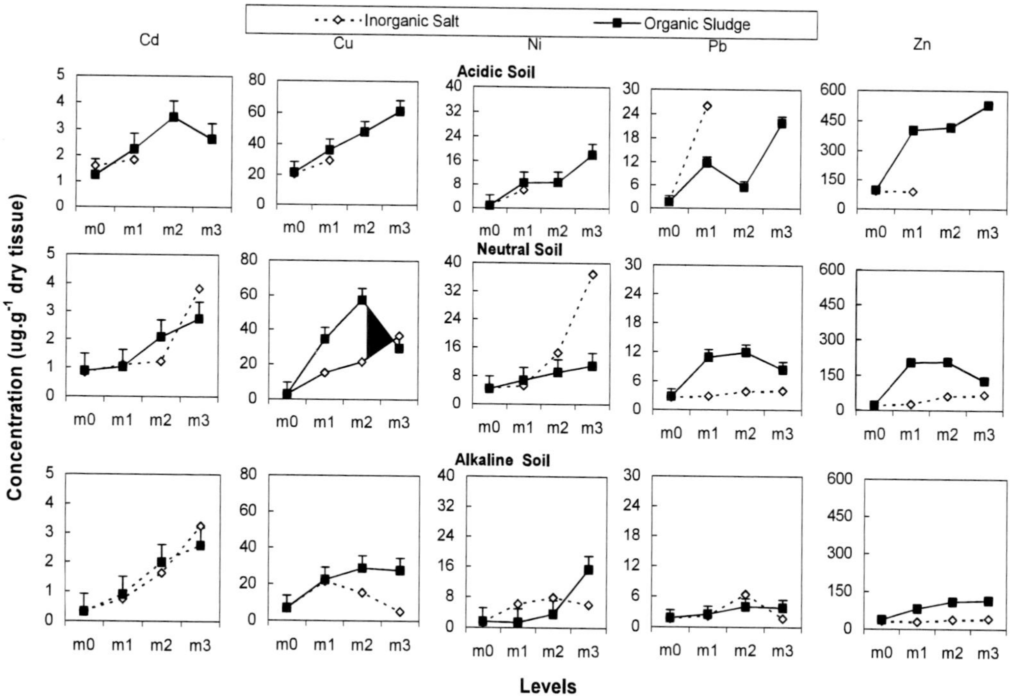

Fig. 8. Effect of inorganic salts and organic sludge on concentration of heavy metals in spinach plant.

Irrespective of the metal source, most of the heavy metals are either required in very low concentrations for normal growth of plants but at higher levels they are often toxic to plants. In the present study, the Cd concentration of spinach tissue was found to be within the acceptable range found in plants under field conditions and was below phytotoxic level (Cd 5-10 mg kg^{-1}) reported by MacLean et al. (1987) except in inorganically polluted acidic soil at m_2 and m_3 levels of metal addition (Fig. 8). Heavy metals supplied through sludge increased the total Cd uptake 2.6 times as compared to inorganic source (Table 6). Different concentration levels (m_1, m_2, m_3) also increased Cd uptake significantly by 70, 170 and 80 per cent over control (m_0), respectively. Root and leafy vegetables are known to accumulate Cd from soil as compared to other heavy metals (Williams and David, 1973; Srikanth and Reddy, 1991). Generally, the Cd concentrations in the plants ranged below 1mg kg^{-1} and could be considered as "normal". Low contents and the neutral to slightly alkaline pH (6.5-7.8) in the soil reduce the mobility and the uptake of Cd.

Table 7. Effect of heavy metals added through inorganic salts or enriched acidic sludge on uptake of Cu by spinach tissue (µg/plant)

Cd	Neutral soil			Alkaline soil			Acidic soil			Mean		
	I	O	Av	I	O	Av	I	O	Av	I	O	Av.
m_0	0.11	0.16	0.13	0.04	0.04	0.04	0.14	0.12	0.13	0.10	0.10	0.10
m_1	0.17	0.34	0.26	0.06	0.08	0.07	0.10	0.26	0.18	0.11	0.23	0.17
m_2	0.17	0.72	0.44	0.10	0.20	0.15	0.01	0.42	0.22	0.09	0.45	0.27
m_3	0.24	0.35	0.30	0.10	0.34	0.22	0.01	0.07	0.04	0.12	0.25	0.18
Av	0.17	0.39	0.28	0.07	0.17	0.12	0.06	0.21	0.14	0.10	0.26	
Effects	Soil(Sl)	Sludge (Sd)	Conc.	Sl x Sd	Sd xC	Sl x C	Sl x Sd x C					
SEm	0.01	0.01	0.01	0.01	0.01	0.01	0.02					
CD (p=0.05)	0.03	0.03	0.03	0.03	0.03	0.03	0.06					

Table 8. Effect of heavy metals added through inorganic salts or enriched acidic sludge on uptake of Ni by spinach tissue (µg/plant)

Pb	Neutral soil			Alkaline soil			Acidic soil			Mean		
	I	O	Av	I	O	Av	I	O	Av	I	O	Av.
m_0	0.33	0.49	0.41	0.22	0.21	0.21	0.15	0.15	0.15	0.23	0.28	0.26
m_1	0.43	3.68	2.06	0.17	0.23	0.20	1.37	1.34	1.36	0.66	1.75	1.20
m_2	0.54	4.11	2.33	0.38	0.41	0.40	0.01	0.66	0.34	0.31	1.73	1.02
m_3	0.26	1.09	0.67	0.05	0.49	0.27	0.01	0.54	0.28	0.11	0.71	0.41
Av	0.39	2.34	1.37	0.20	0.34	0.27	0.39	0.67	0.53	0.33	1.12	
Effects	Soil(Sl)	Sludge (Sd)	Conc.	Sl x Sd	Sd x C	Sl x C	Sl x Sd x C					
SEm	0.02	0.02	0.03	0.03	0.04	0.04	0.06					
CD (p=0.05)	0.06	0.06	0.09	0.09	0.11	0.11	0.17					

The mean concentration of Cu in spinach tissue in all soils at higher levels (m_2 and m_3) of metal application especially, through sludge exceeded the limit prescribed by Allaway (1968). This might be due to release of Cu from organic matter during the course of microbial decomposition and higher solution concentration could be favoured by lower pH value of acidic soil. Such levels could cause toxicity to human beings through continued consumption of such plants. Heavy metals supplied through sludge increased the total Cu uptake approximately four times as compared to inorganic source (Table 7). Different concentration levels (m_1, m_2, m_3) increased total Cu uptake significantly over control (m_0) by 266.0, 411.3 and 79.2 per cent, respectively. Copper is an essential nutrient for plants and animals. However, the metal may become toxic to plants at tissue concentration higher than 20 mg kg^{-1}.

Table 9. Effect of heavy metals added through inorganic salts or enriched acidic sludge on uptake of Pb by spinach tissue (μg/plant)

Ni	Neutral soil			Alkaline soil			Acidic soil			Mean		
	I	O	Av	I	O	Av	I	O	Av	I	O	Av.
m_0	0.58	0.77	0.67	0.16	0.18	0.17	0.07	0.07	0.07	0.27	0.34	0.31
m_1	0.80	2.28	1.54	0.48	0.12	0.30	0.32	0.98	0.65	0.53	1.12	0.83
m_2	2.02	3.12	2.57	0.47	0.36	0.42	0.01	1.06	0.54	0.83	1.51	1.17
m_3	2.35	1.41	1.88	0.18	2.01	1.10	0.01	0.44	0.23	0.85	1.29	1.07
Av	1.44	1.89	1.67	0.32	0.67	0.50	0.10	0.64	0.37	0.62	1.07	
Effects	Soil(Sl)	Sludge (Sd)	Conc.	Sl x Sd	Sd x C	Sl x C	Sl x Sd x C					
SEm	0.03	0.02	0.03	0.04	0.06	0.05	0.08					
CD (p=0.05)	0.09	0.06	0.09	NS	0.17	0.14	0.23					

Table 10. Effect of heavy metals added through inorganic salts or enriched acidic sludge on uptake of Zn by spinach tissue (μg/plant)

Zn	Neutral soil			Alkaline soil			Acidic soil			Mean		
	I	O	Av	I	O	Av	I	O	Av	I	O	Av.
m_0	2.61	3.63	3.12	4.13	4.58	4.35	8.06	8.98	8.52	4.93	5.73	5.33
m_1	3.93	69.08	36.50	2.27	7.33	4.80	4.60	46.14	25.37	3.60	40.85	22.22
m_2	8.21	71.59	39.90	2.23	11.12	6.68	0.01	51.09	25.37	3.48	44.60	24.04
m_3	4.17	16.32	10.24	1.22	14.94	8.08	0.01	13.15	6.58	1.80	14.80	8.30
Av	4.73	40.16	22.44	2.46	9.49	5.98	3.17	29.84	16.51	3.45	26.50	
Effects	Soil(Sl)	Sludge(Sd)	Conc.	Sl x Sd	Sd x C	Sl x C	Sl x Sd x C					
SEm	0.41	0.33	0.47	0.58	0.81	0.66	1.15					
CD (p=0.05)	1.17	0.94	1.34	1.65	2.31	1.88	3.27					

Nickel is a naturally occurring element present in soil, water and air and other biological materials. Nickel is required at low concentration for vital functions but is toxic at high concentration. In plants, patchy discoloration, premature senescence, yellowing of old leaves and growth reduction are common symptoms of nickel toxicity (Scott-Fortsamand, 1977). Normal tissue content of Ni is 1mg kg^{-1} dry weight (Allaway, 1968). The mean Ni content of spinach tissue grown in pot experiment was found to be below phytotoxic range (>50 mg kg^{-1}) (Fig. 8). Heavy metals supplied through sludge increased the total Ni uptake by 1.73 times as compared to inorganic source (Table 8). Different concentration levels (m_1, m_2, m_3) also increased total Ni uptake significantly over control (m_0) by 167.7, 277.4 and 245.2 per cent, respectively.

Lead is not reported to be essential for plants and animals and higher levels could be toxic. The normal range of Pb found in plants is 0.1 to 10 mg kg^{-1} (Allaway, 1968). Even though lead is considered to be an immobile element and tightly bound in roots (Malone et al., 1974), its concentration was found to be reaching up to 21.8 and 26 mg kg^{-1} and above in spinach tissue grown on sludge (at m_3 level) or inorganically (at m_1, m_2 and m_3 levels) polluted acidic soil (Fig. 8). Heavy metals supplied through sludge increased the total Pb uptake of spinach 3.4 times as compared to inorganic source (Table 9). Different levels (m_1,

m_2, m_3) increased total Pb uptake significantly by 316.5, 292.3 and 57.7 per cent over control (m_0), respectively.

Though essential for plants, Zn can be toxic if the tissue concentration exceeds 200 mg kg^{-1} level (Allaway, 1968). However, the content of zinc in spinach grown on polluted alkaline soil was within the safe limits for animal and human consumption (Fig. 8). In neutral and acidic soil, it was present in toxic concentrations at all levels of sludge addition except for m_3 level in neutral soil. The strong growth depression observed at higher levels (m_2 and m_3) of metal application probably caused by the high levels of Zn. Heavy metals supplied through sludge increased total Zn uptake of spinach more than seven fold as compared to inorganic salt (Fig. 8). Different levels (m_1, m_2, m_3) increased total Zn uptake significantly by 316.9, 351.0 and 55.72 per cent over control (m_0), respectively (Table 10).

Relationships Between the Concentrations, Total Uptake of Heavy Metals in Spinach and the Intensities of Different Chemical Fractions of Heavy Metals in Polluted Soils and Soil pH

Simple correlation coefficients (r) computed between the concentrations, total uptake of heavy metals in spinach tissue and different chemical fractions of respective heavy metals are presented in Table 11.

Table 11. Simple correlation coefficients between the concentrations, total uptake of heavy metals in spinach and different chemical fractions of heavy metals in polluted soils and soil pH

Heavy metals	Varia-bles	Chemical fractions of heavy metals					Total content	Soil pH	1M $CaCl_2$	0.005M DTPA
		(F_1)	(F_2)	(F_3)	(F_4)	(F_5)				
Cd	Conc. in plant	0.704**	0.716**	0.750**	0.516*	0.559**	0.744**	-0.546**	0.810**	0.804**
	Total uptake	0.310	0.214	0.250	-0.012	0.250	0.350	-0.447*	0.394	0.378
Cu	Conc. in plant	0.523*	0.568**	0.650**	0.624**	0.741**	0.674**	-0.507*	0.752**	0.762**
	Total uptake	0.297	0.190	0.323	0.295	0.484*	0.373	-0.233	0.251	0.502*
Ni	Conc. in plant	0.656**	0.851**	0.866**	0.926**	0.968**	0.634**	-0.551**	0.861**	0.924**
	Total uptake	0.410	0.192	0.366	0.493*	0.519*	0.492*	-0.492*	0.531*	0.610**
Pb	Conc. in plant	0.086	0.284	0.156	0.352	0.393	0.399	-0.422	0.422	0.095
	Total uptake	-0.109	0.139	-0.034	0.062	0.219	0.178	-0.263	0.088	0.237
Zn	Conc. in plant	0.834**	0.625**	0.658**	0.576**	0.294	0.504*	-0.396	-0.011	0.635**
	Total uptake	0.386	0.467*	0.553**	0.412	0.041	0.273	-0.300	-0.086	0.527*

The concentration of Cd in spinach tissue was significantly and positively correlated with total Cd content of soil and all chemical fractions of Cd in soil. However, total Cd uptake by spinach failed to show statistically significant correlation with any of the Cd fraction. The concentration of Cd in spinach tissue was found to significantly correlated with 1M $CaCl_2$ and 0.005M DTPA fractions. Street et al. (1978) also noted an excellent correlation with Cd-concentration in several plant species and DTPA extractable Cd. These results revealed that DTPA extractable Cd was a good indicator of the bioavailable fraction of this metal in soil.

The concentration of Cu in spinach tissue showed significant and positive correlation with total Cu content of soil and all chemical fractions of Cu in soil. Total Cu uptake by spinach had significant ($p=0.05$) positive correlation with residual Cu fraction in soil. The concentration of Cu in spinach tissue showed significant and positive correlation with 1M $CaCl_2$ and 0.005M DTPA fractions. Total Cu uptake was significantly correlated with 0.005M DTPA fraction at $p = 0.05$.

The concentration of Ni in spinach tissue was significantly and positively correlated with total Ni content of soil and all chemical fractions of Ni in soil. Total Ni uptake by spinach showed significant positive correlation with iron and manganese oxide bound and residual Ni fractions at $p = 0.05$ and 0.01, respectively. The concentration of Ni also found to be significantly and positively correlated with 1M $CaCl_2$ and 0.005M DTPA fractions. Total Ni uptake was significantly correlated with 1M $CaCl_2$ and 0.005M DTPA fraction at $p = 0.05$ and 0.01, respectively.

Neither the concentration of Pb nor the total Pb uptake of spinach showed any statistically significant correlation with total Pb content in soil or any chemical fraction of Pb in soil.

The concentration of Zn in spinach tissue was significantly and positively correlated with total Zn content of soil and all chemical fractions of Zn in soil except residual fraction. Total Zn uptake by spinach showed significant and positive correlation with NaOAc extractable and organically bound Zn fractions at $p = 0.05$ and 0.01, respectively. The concentration of Zn in spinach tissue as well as total uptake was significantly and positively correlated with 0.005M DTPA fraction at $p = 0.01$ and 0.05, respectively.

Good correlation between plant uptake and DTPA-extractable content of heavy metals that has been reported in the past also by Lindsay (1972) and He and Singh (1994).

The concentration of Cu, Cd and Ni in spinach tissue showed significant negative correlation with soil pH, however, the total uptake of Cd and Ni by spinach had a significant (at $p=0.05$) and negative correlation with soil pH. Valdares et al. (1983) and Narwal et al. (1983) also reported a significant and negative correlation between total heavy metal uptake by plants and pH. They reported that uptake of Ni and Zn by plants was significantly larger in the acidic soil of lower pH value than calcareous soil having higher soil pH. A decrease in the metal concentration with increased pH has been reported for various plant species (John and van Laerhoven, 1976; Andersson and Nilsson, 1976; Street et al., 1978; Narwal et al. 1983), and has been associated with increased adsorption of heavy metals by soil components, strong complexing with organic matter, and precipitation of minerals containing these heavy metals.

CONCLUSION

The highest dry yield of spinach was recorded in neutral soil followed by alkaline and acidic soils. Dry yield was higher with enriched sludge as compared to inorganic salt and the fresh and dry yield of spinach was drastically reduced at the highest level of heavy metals supplied through inorganic salt. The effect was more pronounced in acidic soil. The highest uptake of all heavy metals by spinach was recorded in neutral soil. Total uptake of heavy metals through enriched sludge treatment was manifold higher than inorganic salt treatment.

Residual fraction was the most abundant pool among the all chemical fractions in untreated soils. The greater percentage of all heavy metals except Cd was present in Fe-Mn oxide bound and residual fractions. In case of Cd, NaOAc extractable and organically bound fractions were the abundant pools. As compared to inorganic salts, sludge application decreased the residual fraction of Cd, Cu, Ni and Zn but increased it for Pb. Sludge application increased water soluble and exchangeable Zn, NaOAc extractable Cu, Ni and Zn, organically bound Cd, Cu, Ni and Zn, iron and manganese oxide bound Ni and Zn but decreased water soluble and exchangeable Cd, Cu, Ni and Pb, NaOAc extractable Cd and iron and manganese oxide bound Cd and Pb.

Water soluble and exchangeable Cd, Cu, Ni and Zn and residual Pb showed significant negative correlation with the soil pH. The 1M $CaCl_2$ extractable amount of Cd, Cu and Ni showed significant positive correlation with the concentration of these heavy metals in spinach tissue. The DTPA extractable Cd, Cu, Ni and Zn showed significant positive correlation with the concentration of these heavy metals in spinach tissue. 0.005M DTPA extractable Ni and Zn also showed significant positive correlation with total uptake of these metals by spinach plant.

Thus, sludge application promoted dry matter production of spinach in all soils; however, higher level of heavy metals in it may pose health hazards as well. The DTPA extractable heavy metal content of soils can serve a suitable index of their bioavailability to plants. The present studies also revealed that care has to be taken while applying industrial sludge containing heavy metals as a substitute of organic manures in agricultural fields, especially in acidic soil.

REFERENCES

Adriano, D.C. 1986. Trace elements in terrestrial environment. Springer-Verlag, New York.

Ahnstrom. Z.S. and Parker, D.R. 1999. Development and assessment of a sequential extraction procedure for the fractionation of soil cadmium. *Soil Sci. Soc. Am. J.* 63 : 1650-1658.

Allaway, W.H. 1968. Agronomic controls over environmental cycling of trace elements. *Advan. Agron.* 20 : 235-279.

Andersson, A. and Nilsson, K.O. 1976. Influence on the levels of heavy metals in soil and plant from sewage sludge used as fertilizer. *J. Agric. Res.* 6 : 151-159.

Baldwin, A., Brown, T.A., Beckett, P.H.T. and Elliot, G.E.P. 1983. The forms of combination of Cu and Zn in digested sewage sludge. *Water Res.* 17:1935-1944.

Bell, P.F., James, B.R. and Chaney, R.L. 1991. Heavy metal extractability in long-term sewage sludge and metal salt amended soils. *J. Environ. Qual.* 20 : 481-486.

Chang, A.C., Page, A.L., Warneke, J.E., Resketo, M.R. and James, T.E. 1983. Accumulation of cadmium and zinc in barley grown on sludge treated soils. *J. Environ. Qual.*12 : 391-397.

Chapman, H, D. 1969. Cation exchange capacity. *In* : Methods of Soil Analysis. C.A. Black (ed.). Part 2. Agron. Monogr. 9. U.S.A., Madison, WI. pp. 894-897.

Chlopecka, A. 1993. Forms of trace metals from inorganic sources in soils and amounts found in spring barley. *Water, Air and Soil Pollution.* 69 : 127-134.

Cunningham, J.D., Keeney, D.R. and Ryan, J.A. 1975. Phytotoxicity and uptake of metals added to soils as inorganic salts or in sewage sludge. *J. Environ. Qual.* 4 : 455-459.

Dowdy, R.H. and Larson, W.E. 1975. Metal uptake by barley seedling grown on soils amended with sewage sludge. *J. Environ. Qual.* 4:229-233.

Dudka, S., Ponce Hernandez, R., Tate, G. and Hutchinson, T.C. 1996. Forms of Cu, Ni and Zn in soils of sudbury, Ontario and the metal concentration in plants. *Water, Air and Soil Pollution.* 90 : 531-542.

Emmerich, W.E., Lund, L.J., Page, A.L. and Chang, A.C. 1982. Movement of heavy metals in sewage sludge treated soils. *J. Environ. Qual.*11 : 174-178.

Gupta, S.K. and Chen, K.Y. 1975. Partitioning of trace metals in selective chemical fractions of near shore sediments. *Environ. Lett.* 10 : 129-158.

Haghiri, 1974. Plant uptake of cadmium as influenced by cation exchange capacity, organic matter, zinc and soil temperature. *J. Environ. Qual.* 3 : 180-183.

He, Q.B. and Singh, B.R. 1994. Crop uptake of cadmium from phosphorous fertilizers : II relationship with extractable cadmium. *Water, Air and Soil Pollution.* 74 : 267-280.

John, M.K. and van Laerhoven, C.S. 1976. Effects of sewage composition, applicataion rates and lime regime on plant availability of heavy metals. *J. Environ. Qual.* 5 : 246-251.

John, Matt K., Vanlarhoven, Cornelis, J. and Chuah, Hong H. 1972. Factors affecting plant uptake and phytotoxicity of cadmium added to soils. *Environ. Sci. Technol.* 6: 1005-1009.

Jordao, C.P. and Nickless, G. 1989. Chemical association of Zn, Cd, Pb and Cu in soils and sediments determined by the sequential extraction technique. *Environ. Technol. Lett.* 10 : 743-752.

Kuo, S., Heilman, P.E. and Baker, A.S. 1983. Distribution and forms of copper, zinc, cadmium, iron and manganese in soils near a copper smelter. *Soil Sci.* 135 : 101-109.

LeClaire, J.P., Chang, A.C., Levessque, C.S. and sposito, G. 1984. Trace metal chemistry in arid zone field soils amended with sewage sludge. IV Correlations between zinc uptake and extracted soil zinc fractions. *Soil Sci. Soc. Am. J*.48 : 509-513.

Lindsay, W.L. 1972. Zinc in soil and plant nutrition. *Adv. Agon.* 24 : 148-186.

Lindsay, W.L. and Norvell, W.A. 1978. Development of a DTPA soil test for zinc, iron, manganese and copper. *Soil Sci. Soc. Am. J.* 42:421-428.

Lund, L.J., Page, A.L. and Nelson, C.O. 1976. Movement of heavy metals below sewage disposal ponds. *J. Environ. Qual.* 5:330-334.

MacLean, K.S., Robinson, A.R. and Macconnell, H.M. 1987.The effect of sewage sludge on the heavy metal content of soils and plants tissue. *Commun. Soil Sci. Plant Anal.* 18: 1303-1316.

Mahler, R.J., Bingham, F.T. and Page, A.L. 1978. Cadmium enriched sewage sludge application to acid and calcareous soils: effect on yield and cadmium uptake by lettuce chard. *J. Environ. Qual.* 7: 274-281.

Malone, C.D., Koepp, D.E. and Miller, R.J. 1974. Location of lead accumulation by corn crops. *Plant Physiol.* 53:388-394.

Ma, Lena Q. and Rao, Gade N. 1997. Chemical fractionation of cadmium, copper, nickel and zinc in contaminated soil. *J. Environ. Qual.* 26 : 259-264.

Morera, M.T., Echeverria, J.C. and Garrido, J.J. 2001. Mobility of heavy metals in soils amended with sewage sludge. *Can. J. Soil Sci.* 81 : 405-414.

Narwal, R.P., Singh, B.R. and Panwar, A.R. 1983. Plant availability of heavy metals in a sludge-treated soil : I. effect of sewage sludge and soil pH on the yield and chemical composition of rape. *J. Environ. Qual.* 12(3) : 358-365.

Page, A.L., Miller, R.H. and Kenney, D.R. 1982. Methods of Soil Analysis-Part 2(Ed) No. 9, agronomy Series ASA-SSSA Publisher, Madison, Wisconsin, U.S.A.

Payne, G.G., Martens, D.C., Winarko, C. and Perera, N.F. 1988. Form and availability of copper and zinc following long term copper sulfate and zinc sulfate applications. *J. Environ. Qual.* 17 : 707-711.

Pietz, R.I., Peterson, J.R., Hinsely, T.D., Ziegler, Y.E.L., Redborg, K.E. and Luehigh, C. 1983. Sewage sludge application to calcareous strip-mines spoil : II Effect on spoil and corn cadmium, copper, nickel and zinc. *J. Environ. Qual.* 12(4) : 463-467.

Ramachandran and D'Souza, T.S. 1999. Plant uptake of cadmium, zinc and manganese from soils amended with increasing levels of Cd-enriched sewage sludges and city composts. *J. Indian Soc. Soil Sci.* 47(4) : 738-743.

Ramos, L., Hernandez, L.M. and Gonzalez, M. J. 1994. Sequential fractionation of copper, lead, cadmium and zinc in soils from or near Donana National Park. *J. Environ. Qual.* 23(1) : 50-57.

Rudd, T., Sterritt, R.M. and Lester, J.N. 1984. Formation and conditional stability constants of complexes formed between heavy metals and bacterial extracellular polymers. *Water Res.* 18:379-384.

Scott-fordsmand, J.J. 1977. Toxicity of nickel to soil organisms in Denmark. *Rev. Environ. Cont. Toxicol.* 148 : 1-34.

Silviera, D.J. and Sommers, L.E. 1977. Extractability of copper, zinc, cadmium and lead in soils incubated with sewage sludge. *J. Environ. Qual.* 6(1) : 47-52.

Sims, J.T. and Kline, J.J. 1991. Chemical fractionation of plant uptake of heavy metals in soils amended with co-composted sewage sludge. *J. Environ. Qual.* 20 : 387-395.

Soon, Y.K. and Bates, T.E. 1982. Chemical pools of cadmium, nickel and zinc in polluted soils and some preliminary indications of their availability to plants. *J. Soil Sci.* 33 : 477-488.

Sposito, G., Lund, L.J. and Chang, A.C. 1982. Trace metal chemistry in arid zone field amended with sewage sludge: I. Fractionation of Ni, Cu, Zn, Cd and Pb in solid phases. *Soil Sci. Soc. Am. J.* 13 : 372-376.

Srikanth, R., Raja, S. and Reddy, P. 1991. Lead, cadmium and chromium levels in vegetables grown in urban sewage sludge-Hyderabad, India. *Food Chemistry.* 40 : 229-234.

Street J.J., Sabey, B.R. and Lindsay, W.L. 1978. Influence of pH, phosphorous, cadmium, sewage sludge and incubation time on the solubility and plant uptake of cadmium. *J. Environ. Qual.* 7(2) : 286-290.

Taylor, Robert W., Xiu, He, Mehadi, Ahmend A., Shuford, James W. and Tadesse Wubishet. 1995. Fractionation of residual cadmium, copper, nickel, lead and zinc in previously sludge amended soil. *Commun. Soil Sci. Plant Anal.* 26(13-14) : 2193-2204.

Tessier, A., Campbell, P.G.C. and Bission, M. 1979. Sequential extraction procedure for the speciation of particular trace metals. *Anal. Chem.* 51(7) : 841-851.

Valdares, J.M.A.S., Gal, M., Mingelgrin, U. and Page, A.L. 1983. Some heavy metals in soils treated with sewage sludge, their effects on yield and their uptake by plants. *J. Environ. Qual.* 12(1) : 49-57.

Williams, C.H. and David, D.G. 1973. The effect of super phosphate on cadmium content of soil and plant. *Aus. J. Soil Res.* 11 : 43-56.

Xian, X.F. and Shokohifard, G.I. 1989. Effect of pH on chemical forms and plant availability of cadmium, zinc and lead in polluted soils. *Water, Air and Soil Pollution.* 45 : 265-273.

In: Hazardous Materials in the Soil and Atmosphere
Editor: Robert C. Hudson, pp. 117-142

ISBN 1-60021-250-6
© 2006 Nova Science Publishers, Inc.

Chapter 5

REMOVAL OF PHENOLICS FROM AQUEOUS SOLUTIONS BY USING MODIFIED CLAYS AS ADSORBENTS

Saadet Yapar
Ege University, Engineering Faculty, Chemical Engineering Department,
Bornova, İzmir, Turkey

ABSTRACT

Phenolic compounds are priority pollutants due their toxic and harmful effects even at low concentrations. In addition to the biological sources such as olive oil mill effluents and animal wastes; the sources of phenol contamination are mainly the petrochemical, coal conversion and phenol producing industries. There are many methods for the removal of phenol from aqueous solutions, adsorption seems to offer a better prospect than the others, which require high pressure and /or temperature, and complex equipment. Since synthetic adsorbents are relatively expensive, the use of natural adsorbents is an active area of research. Additionally, some biological materials and industrial wastes are also used in the removal of phenol. Among the naturally occurring adsorbents, clay has the most common application due to its outstanding adsorption properties. However, the modified forms of clay are generally used in the removal of phenolics since the negatively charged and hydrophilic surfaces of the mineral render it to be less efficient. The modified forms are obtained through calcinations, pillaring, acid activation, and the replacement of exchange cations with quaternary alkylammonium derivatives. The adsorption properties of the modified forms are generally tested by batch experiments. Application of the results to industrial scale requires the determination of kinetic and isotherm models. Langmuir and Freundlich models are widely used in the determination of the adsorption behavior of phenolics on modified surfaces. In terms of design considerations, the model proposed should be as simple as possible and contain minimal parameters. This article will cover a literature review on the removal of phenolics by using modified clays. The methods used in the modification and the adsorption properties of the modified forms will be summarized. The forms of isotherms and isotherm equations used in the modeling of adsorption behavior will also be given.

1. INTRODUCTION

1.1. Phenolic Compounds and Their Environmental Impacts

Although it is found in animal wastes and organic material, phenol is mainly a man-made chemical. It is produced in large amounts, mostly as an intermediate in the production of other chemicals. It is a colorless or white solid with a characteristic sweet tarlike odour. Phenol has the highest water solubility among its derivatives. It is also very soluble in alcohol, chloroform, ether, glycerol, carbon disulphide, petrolatum, volatile and fixed oils and aqueous alkali hydroxides. Since the aromatic ring is the dominant character, many of the phenolics have almost the same physical properties as phenol. Due to this reason, phenol is used as a model pollutant in environmental research.

In addition to being a disinfectant, that is the main area of use; phenol is also used as a slimicide, as a reagent in chemical analysis and for the manufacture of artificial resins, medical and industrial organic compounds and dyes. It is an important raw material in the petrochemical, pharmaceutical, plastic and pesticides chemical industries. Phenol is largely used as intermediate in the production of phenolic resins and caprolactam.

The standard limits established for the release of phenolic compounds are normally lower than 0.5 mg/L but the content of phenolic compounds in industrial wastewaters changes to the range of 200 to 2000 mg/L [1]. In the case of single or small releases, the half-life of phenol in water is between 2 to 20 days, it is long in comparison to its half-life in air and soil. If a large amount is released at one time or if a steady amount is released over a long time, the half-life will increase. Therefore, the surface waters and surrounding air nearby the above mentioned industries would be more contaminated. The other main sources of phenol contamination are oil refinery discharges, coal conversion plants and municipal waste treatment plant discharges or spills. Potential sources of phenol contamination especially in air are agricultural chemicals, disinfectants, general antibacterials and antiseptics, household hard surface cleaners, lubricating oils, automotive chemicals, paint and varnish removers, pharmaceutical preparations and wood office work surfaces.

Among the derivatives of phenol, chlorophenols are widely used as agricultural chemicals, pharmaceuticals, biocides, and dyes. They are produced by the substitution of the aromatic ring of phenol with one to five chlorine atoms. Although there are 19 chlorophenol derivatives, 2-chloro-, 4-chloro-, 2,4-dichloro-, 2,6-dichloro-, 2,4,6-trichloro-, and pentachorophenol are the most important derivatives [2]. They are slightly soluble in water and highly soluble in alcohols, the other physicochemical properties are given in Table 1. Besides their strong characteristic odors, they also impart a strong disagreeable taste to drinking water.

The effects of phenol on humans are; irritation of the nose, throat and eyes, and skin burns. The result of exposure to high levels is acute poisoning, causing oscillation in the respiration rate, decrease in body temperature, cyanosis, muscular weakness, weak or occasionally rapid pulse and coma. Chronic exposure causes systemic problems including vertigo, digestive difficulties, skin eruptions, nervous problems and headaches. Although chronic effects of human exposure to low doses of chlorophenols have not been reported, the toxic doses of pentachorophenol cause severe pathological changes in the gastrointestinal tract, lungs, spleen, liver, kidneys and cardiovascular system.

Table 1. Physicochemical Properties of Phenol and Chlorophenols

Compound	Solubility, g/L	pK_a	log K_{ow}
Phenol	82	9.89	1.57
2-Chlorophenol	28.5	8.52	2.29
4-Chlorophenol	27.1	9.37	2.39
2, 4-Dichlorophenol	4.5	7.9	2.9
2, 4, 6- Trichlorophenol	0.8	5.99	3.72
Pentachlorophenol	0.014	4.74	5.25

*log K_{ow}, logarithm of compounds partition coefficient between octanol and water

Phenol and its derivatives have a severe environmental impact. They are toxic and harmful even at low concentrations. Acute toxic effects of phenol include the death of animals, birds or fish and death or low growth rate in plants. Additionally, pregnant animals that drink water contaminated with phenol give birth to offspring that has low birth weights and birth defects. Shortened lifespan, reproductive problems, lower fertility and changes in appearance or behavior are the long-term effects. The toxicity and bactericidal activities of chlorophenols depend on the degree of chlorination and the position of the chlorine atoms. Although all chlorophenols possess bactericidal activities, it attains a maximum for the trichlorophenols. Enzyme inhibitions of oxidative phosphorylation and the synthesis of adenosine triphosphate are the metabolic effects of chlorophenols. Additionally, they are highly toxic to algae; trichlorophenols are more toxic than dichlorophenols and pentachlorophenol is used as an algicide. They also have phytotoxicity; the terrestrial and aquatic plants can in certain cases adsorb, transform and eliminate chlorophenols without harm, however most of ten these plants are sensitive to the phytotoxicity of chlorophenols.

1.2. Adsorption and Adsorbents

Since the adsorption and absorption are homophonic words, beginners generally confuse the process defined by these words. Adsorption means to take up the molecules and hold on the surface of a solid or liquid whereas, absorption is to suck up or take in like a sponge. Adsorption often occurs between solid and liquids and the components of adsorption are described as adsorbent, solid and adsorbate, solute.

Because the adsorption is principally governed by the interactions between the solid surface and solute, the area and charge of the adsorbent surface are the parameters determining the adsorption efficiency. Thus, uniform surface properties, a large surface area and an opposite charge with the adsorbate are of utmost importance in the selection of the adsorbent to be used in wastewater treatment. In addition to the charge and surface area, the cost of unused adsorbent and the regeneration of the exhausted adsorbent are the other parameters making an adsorbent to be cost effective.

The adsorbents used in the removal of phenol show a large diversity. Activated carbon with uniform surface properties and large surface areas is the best among them. However, the use of activated carbon for industrial wastewater treatment is a capital-intensive process having the problems such as regeneration of activated carbon, intraparticle resistance in the adsorption process in practice, cost of manufacture and blinding of the macroporous spaces by large molecules [1, 3]. Depending on the source, wastewater will contain a large variety of pollutants with changing molecular weights. For instance, phenol and acetone are produced by the cumene oxidation process. The effluent of the process contains 2-3% phenol, 3-6% acetone, up to 0.1% aromatic hydrocarbons, mainly cumene and α-methylstyrene, and 2-3% sodium salts, mainly formate and sulphate [4]. The effluents of the petrochemical industry and olive oil mills contain oil and large molecules such as polyalcohols and polyphenols [3, 5]. These molecules will blind the macroporous space on the activated carbon outer surface and thus the microporous space will be ineffective. In some applications, the activated carbon is used in combination with other adsorbents such as organo clays, which are capable of adsorbing large molecules, to overcome this problem. The other solution is to produce the activated carbon with micropores and interconnecting mesopores. It is reported that the activated carbon obtained from waste tires have the mesoporous structure [6]. It has a phenol adsorption capacity that is slightly lower than that of commercial activated carbon but it could not be perfectly regenerated.

Both the removal of laden pollutant from the adsorbent and the protection of adsorption properties are the main goals of the regeneration processes. Although solvent regeneration is possible, exhausted carbons are generally thermally regenerated but 5-10% of the carbon is typically lost by attrition, which is common for both types of regeneration and burnoff during each cycle [6]. The cost of the required fuel is another parameter increasing the operational cost.

Because of the above-mentioned issues, an extensive research has been conducted to develop innovative and promising adsorbents to utilize in the treatment of chemically contaminated waters. In addition to the naturally occurring adsorbents, many researchers have also investigated the adsorptive properties of unconventional adsorbents, such as polyurethane foam, olive seeds, peat, fly ash, perlite, coal, and rice husk. All these low-cost potential adsorbents are classified into three groups: industrial wastes, biological materials and minerals. They all have high porosity and large surface areas but these properties cannot be sufficient to obtain high efficiencies in the removal of phenolic compounds. The interactions between surface and phenol molecules and pore size should also be considered. The diameter of a phenol molecule is about 6Å [7] and therefore the adsorbents should have a pore size, which is large enough for the entrance of the phenol molecules. Applying the pretreatment to remove the impurities in the pores of some adsorbents can vary the pore size. Coal is an example for such adsorbents, it posses good microporosity and therefore a large surface area [1]. The different coal conversion processes such as oxidation, pyrolysis, and solvent extraction are applied to improve the surface porosity and thus the adsorption properties. Ahmaruzzaman and Sharma [1] have reported that the adsorption capacity of residual coal increased after treatment with phosphoric acid and was about 48% as efficient as standard activated carbon.

INDUSTRIAL WASTES

The increasing amounts of industrial wastes in parallel with industrialization create negative environmental impacts eventhough they do not have any toxic effect on the environment. For this reason, the use of industrial wastes as adsorbents is considered a doubly effective solution for environmental pollution. However, the waste used should have certain structural properties and at least a high surface area. Depending on the nature of the pollutant to be removed some additional properties such as, the pore size and surface charge, will gain importance.

Fly ash is a commonly used industrial waste in adsorption studies [8- 11]. It is generated in huge amounts from various sources such as coal -fired thermal power plants and bagasse-fired boilers of cane sugar mills. Depending on the source, the pore size and surface area and therefore the adsorptive properties of the material will change. Actually, the pore size determines the dimension of the molecule to be adsorbed. For example, fly ash from old power plants has a significant capacity for the sorption of low molecular weight organic contaminants [12]. The phenol adsorption capacity of fly ash is comparable to the adsorption capacities of peat and bentonite when it is in contacted with the phenol solution for a period of 5 h with a pH of 5 [8]. In addition to the high carbon content, the presence of silica and alumina make the fly ash an effective adsorbent for phenol and its derivatives. The placement of various substituents, such as chlorine, in different positions makes the phenol molecule less soluble in water. Although this can be considered as a factor leading the more energetic adsorbent-adsorbate interactions relative to adsorbate-water interactios and thus an increase in the adsorbed amounts, the adsorption of these molecules in their neutral molecular forms to the charged surfaces will not be in considerable amounts due to the weak adsorbent-adsorbate interactions. The existence of a carbonaceous layer on the surface will provide a medium increasing these interactions and therefore the adsorption capability of adsorbent. There are various researches confirming the effect of solubility and adsorbate-adsorbent interactions on the adsorption capability of the adsorbents. For example, Jain and coworkers [13] compared to the adsorption of methylphenols on a carbonaceous adsorbent and an inorganic adsorbent obtained from fertilizer and steel industry wastes, respectively. It was reported that the carbonaceous adsorbent adsorbs more methylphenol than the steel industry wastes and it is about 45% as efficient as standard activated charcoal [13].

BIOLOGICAL MATERIALS

The existence of a highly porous structure makes biological materials potential adsorbents. Moreover, many biological materials are used as raw materials in the preparation of activated carbon because of this property. Although, the activated carbon has more pronounced adsorption capacity than that of its raw materials, the cost is sometimes not balanced by the improved adsorption properties due to the problems occurring in regeneration and during the use in real processes. Therefore the direct uses of the biological materials in adsorption gain importance and there are several works reporting the adsorption properties of agricultural by-products [14-16]. The removal of heavy metals is targeted in many of these earlier studies. The adsorption of pollutants by biological materials is termed as bio-sorption and the material used as bio-adsorbents. Activated sludge utilized in biological waste treatment systems is another material that can be used in the bio-sorption. Since these systems are worldwide, the activated sludge is actually an abundant and cheaper bio-sorbent [17]. Aksu and Yener studied the bio-sorption of phenol and its monochlorinated derivatives using dried sludge [18]. They reported that the potential of dried sludge for the removal of all these pollutants was considerable and it was possible to remove phenol and monochlorophenols at a wide concentration range up to 500 mg/L. The microorganisms in a conventional activated sludge system can only deal with chemical wastewater containing the phenolic compounds in concentrations less than 100 mg/L. Under these considerations, the use of dried activated sludge seems to be an alternative to the most costly materials. Beside the activated sludge, the various researchers investigated the use of fungal biomass [19, 20]. Rao and Viraraghavan [19] have used pretreated non-viable, Aspergillus niger biomass for the removal of phenol and reported that the maximum removal was obtained at an initial pH of 5.1 for the biomass treated with sulfuric acid. They concluded that the adsorption was very strong since only 5% of adsorbed phenol was desorbed. Wu and Yu [20] have compared the effect of varied number of chlorine atoms in the phenol molecule on the bio-sorption by fungal mycelia and they found that the adsorption increased with the increasing number of chlorine atoms. This behavior was attributed to the dominance of partitioning [20].

MINERALS

Both synthetic and naturally occurring minerals such as zeolite, and clay are used as adsorbents. A broad range of pore size and siliceous surfaces are the characteristics for many of them. Zeolite and clays are the most abundant naturally occurring aluminosilicate minerals with negatively charged surfaces. Although they can be synthesized, the natural forms are commonly used as adsorbents. As shown in Table 2, the same elements are included in their chemical compositions but the crystal structures of the minerals are quite different. Clay minerals are a group of magnesium or alumino-phyllosilicates that are composed of octahedral and tetrahedral sheets. Depending on the combinations of octahedral and tetrahedral sheets, clay minerals are grouped as 1:1, 2:1 and 2:1:1 type clays. Among the clay minerals, mostly bentonite, which is a member of 2:1 type smectite clays, is used as an adsorbent. It exhibits a swelling behavior resulting from the weak attraction between the oxygen on the bottom and top of the tetrahedral sheets [21]. Clinoptilolite is the most abundant natural zeolite having a cage like structure with the largest aperture measuring 4.4 by 7.2Å [22]. In contrast to bentonite and other clay minerals, it is free of shrink- swell behavior.

Table 2. The Typical Chemical Compositions of Bentonite and Clinoptilolite

Constituent	Value	
	Bentonite	**Clinoptilolite**
SiO_2	56.63	78.05
Al_2O_3	23.25	6.34
Fe_2O_3	4.41	0.45
Na_2O	2.85	2.57
MgO	1.77	0.33
K_2O	0.48	1.82
CaO	0.47	2.31

Perlite is another natural mineral that can be used as adsorbent. Koumano and Peeva-Antova [23] investigated the adsorption of p-chlorophenol on yellow bentonite and expanded perlite from aqueous solutions. They reported that the rate of adsorption of p-chlorophenol was influenced by the initial concentration of the adsorbate and perlite has a lower adsorption capacity [23].

Although phenol is an ionizable organic compound, it is mainly in the neutral molecular form at pH values 3-8, at higher pH values it dissociates to phenolate and a proton according to the following reaction.

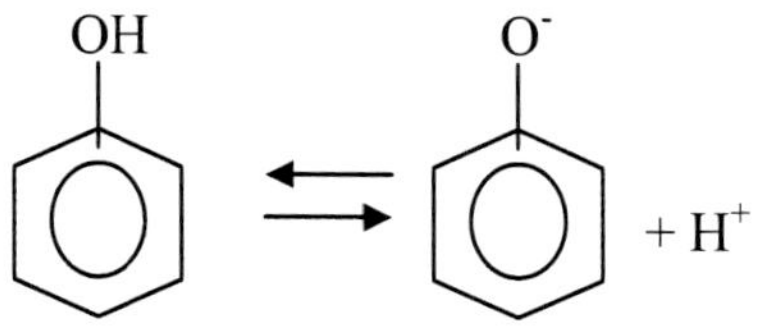

The presence of a negative charge on the mineral surfaces causes a competition between phenol and water molecules in the adsorption from aqueous solutions. Thus, adsorbents with hydrophobic surfaces are the most advantageous for the selective adsorption of phenol from aqueous solutions [24]. Zeolites having a high Si/Al ratio are hydrophobic and this ratio can easily be arranged in synthetic zeolites. Roostaei and Tezel [25] studied the effect of surface hydrophobicity on the adsorption of phenol. They used activated carbon and powder adsorbents such as activated alumina, silicagel, HiSiv 1000 and HiSiv 3000. Since the activated alumina and silicagel are also desiccants, they have a higher affinity to water molecules than phenol. As expected there was no significant adsorption of phenol by silicagel and activated alumina. HiSiv 3000 and 1000 have a zeolite ZSM-5 and zeolite-Y structure, respectively. HiSiv 1000 had the highest adsorption rate among the adsorbents studied. More detailed studies were carried out using this adsorbent and it was found that the adsorption capacity was not concentration dependent and did not change after a 14-regeneration cycle. A similar research was conducted by Khalid et al. [24], they used HFAU zeolites with the Si/Al ratios ranging from 5-100, silicalite with Si/Al=500 and a BEA zeolite with Si/Al=∞. It was concluded that the main parameter defining the adsorption properties of a zeolitic adsorbent was its hydrophobic character.

Şişmanoğlu and Pura [22] studied the effect of the position of the nitro group in a phenol molecule on the adsorption on a natural zeolite, clinoptilolite. They found that o- and m-nitrophenols were adsorbed at lower amounts than p-nitrophenol. This is probably due to the increased width of the molecule in the case of o- and m- substituents. Since the molecules are mainly captured in the pores, the ratio of pore size to the size of molecule is of importance. Replacing the bigger neutralizing cations with smaller ones can provide a relative increase in the pore size but this increase is not enough to allow the entrance of molecules if they are larger than the largest aperture of the zeolite. Therefore the pore size, apertures and hydrophilic surfaces of natural zeolites are the parameters restricting the adsorption. The existence of negatively charged surfaces permits the alteration of surface chemistry through the treatment with cationic surfactants. However, the surface treatment is limited to the external surface due to the size of the surfactant molecule. Depending on the amount of surfactant used and the external cation exchange capacity, a mono- or double layer is formed on the surface. The degree of surface coverage determines the hydrophobicity of the surface.

In the case of double layer, a hydrophilic positively charged surface is obtained. Li et al. [26] observed that surfactant-modified zeolite is effective in sorbing ionizable organic compounds in addition to nonionic organic compounds (NOCs). They also reported that phenol adsorption by surfactant modified zeolite treated to bilayer coverage increased as the pH increased.

The adsorption properties of bentonites have been known for ages. The positively charged species are adsorbed in considerable amounts and the adsorption is very fast in many cases. However, the negative charge and hydrophilicity of the surfaces resulting from the hydration of neutralizing cations cause less adsorption of NOCs due to the preferential adsorption of water. Through modification with alkylammonium cations, both surfaces can be converted from hydrophilic to hydrophobic and the competition in favor of nonionic compounds.

2. Modified Clays

2.1. Organo-Modified Clays

The aim of modification is not limited with the conversion of the surface from hydrophilic to hydrophobic. It is actually carried out to fulfill the requirements of the applications to take place in a broad range. Depending on the ultimate properties of modified clay and the area of use, 2:1 minerals are modified by using different techniques. Bergaya and Lagaly [27] collected these techniques in 11 main groups. They are: 1) adsorption, 2) ion exchange with inorganic cations and cationic complexes, 3) ion exchange with organic cations, 4) binding of inorganic and organic anions, 5) grafting of organic compounds, 6) reaction with acids, 7) pillaring by different types of poly (hydroxo metal) cations, 8) interlamellar or intraparticle and interparticle polymerization, 9) dehydoxylation and calcinations, 10) delamination and reaggregation of smectitic clay minerals, and 11) physical treatment such as lyophilization, ultrasound and plasma.

Ion exchange with organic cations results in organo-bentonites. The mostly used cations are the quaternary ammonium cations with the forms of $(CH_3)_3NR^+$ and $(CH_3)_3NR_2^+$, where R is either an alkyl or aromatic-hydrocarbon group. The properties of organo-bentonite depend on the size and structure of the cation used. For example, a tetramethylammonium smectite behaves like a zeolite with narrow micropores of rather hydrophilic character whereas long chain cations make the interlayer space hydrophobic [28]. In addition to the change in surface hydrophilicity, the expansion of interlayer space occurs; the orientation and the chain length of cations determine the degree of expansion, which is observed by conducting an XRD analyses. Beside the crystalline structure and composition of the mineral; the value of basal spacing, d_L is determined by this analysis and the difference between the original and modified clay basal spacings is the measure of expansion and the orientation of the molecules at the surface. According to the model of Lagaly and Weiss [29], a 4 Å increase in the basal spacing is indicative to a monolayer and the molecules are oriented parallel to the surface. The orientation on the surface is closely related to the amount of adsorbed cations and the cation exchange capacity of bentonite. Depending on the amount of cation used the formation of a monolayer, a bilayer, a pseudotriple layer and/or an inclined structure is possible. When the amount of cation used is equal to the cation exchange capacity a bilayer is formed. In the other cases, at which the amounts are less than the cation exchange capacity a random interstratification of monolayer and bilayer structures are observed [28, 30].

The modification procedure is relatively simple; it involves the purification and conversion to a single ion form, drying, grinding, modification with alkylammonium cations and final drying steps. After the removal of coarse fraction impurities, the dried powder is digested with concentrated hydrogen peroxides. The next steps are to convert the clay into desired ion form and modify the surface by ion exchange. Since the diffusion of cations to the interlayer space is the rate-determining step, medium temperatures should be preferred in the modification to get relatively high diffusion rates and to complete the synthesis in a short time. The last steps are the separation of the modified clay and drying. The drying conditions affect the structure of alkylammonium cations on the surface; freeze –dried samples are more hydrophilic than those of high temperature dried. The difference in the surface lyophilicity is reflected on the adsorption capacities as shown in Figure 1. A close examination of the figure reveals that the adsorbed amounts nearly doubles for the adsorption on the freeze dried adsorbent at high equilibrium concentrations.

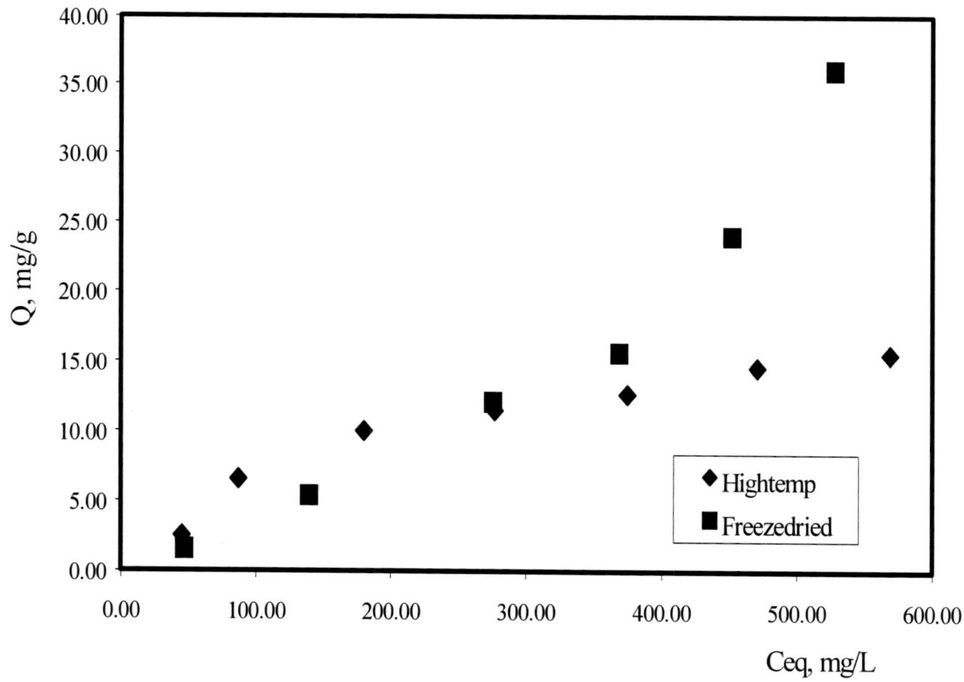

Figure 1. Effect of Drying Conditions on the Adsorbed Amount.

Since the main objective of all studies in the field of adsorption is to get high removal rates, the improvement of adsorption properties is of prime importance. Since this can be achieved by eliminating the factors having negative impacts, elucidating of the adsorption mechanism is important. Many researchers are reported that the magnitude and the mechanism of adsorption are the functions of the molecular structure of exchanged organic cation, the extent of cation exchange, the solution chemistry, the molecular structure of solute and cation exchange capacity of bentonite. The results of the almost innumerable researches on the use of organo - bentonites in the removal of phenolics are also in agreement with these findings. For example, Koh and Dixon [31] studied the adsorption of benzene, phenol and toluene on organo-minerals prepared by using the quaternary ammonium cations with various molecular weights such as Benzyldimethyltetradecylammonium (BDTDA), Hyamine 1622 and Benzyltrimethyl ammonium (BTMA) and three different minerals, montmorillonite, zeolite and sericite, as substrates. It was observed that BDTDA-montmorillonite was the most effective sorbent of the three NOCs and the exchange behaviors of organic cations, the interlayer expansion of organo-montmorillonite, the adsorption capacity of organic cations on montmorillonite and the uptake capacity of aromatic hydrocarbons by organo-montmorillonite were closely related to each other. Juang et al. [32] studied the effect of the interaction between phenols and organo-bentonite and the water solubility of phenols on adsorption capacity. They reported that the adsorption capacity decreased in the order phenol>o-cresol>m-nitrophenol and this order is the same with the order of the water solubility. The relatively high water solubility of phenol results from the strong interactions between phenol and water molecules. The relative magnitudes of the phenol-surface and phenol-water interactions determine the adsorption capacity and therefore the increase in adsorption capacity is parallel to the decrease in solubility.

When a short chain alkyammonium cation is used, the surface will have both hydrophilic and hydrophobic properties, since the considerable surface could not covered by the cation and the hydrophobicity of the surface will be greatly affected by the solvent [33]. The adsorption of phenolic compounds occurs when hydrogen bonds to the water molecules of the inorganic interlayer cation and van der Waals attraction to alkyl chains since the phenol molecule possesses both a localized center of high electron density and a nonpolar aromatic ring [34]. In the case of a long chain alkylammonium cation, the surface properties transform from hydrophilic to hydrophobic, since the organic functional groups of the cation are not strongly hydrated by water [35] and the effect of solvent to the hydrophobicity is not considerable [33]. The alkyl chains of the cation create an organic partition medium for the NOCs and the partition mechanism is closely related to the chemistry of the adsorbate since the partition is conceptually and functionally similar to the dissolution of the adsorbate in a bulk organic solvent such as octanol [36]. Therefore, any structural property causing a decrease in the octanol solubility will create the same effect in the adsorption. In this respect, the sorption of polar organic compounds by partition will be weaker than that of nonpolar compounds. Zhu et al. [35] observed this phenomenon in the adsorption of nitrobenzene and anilin on cetyltrimethylammonium-bentonite. The relative magnitudes of solvent-solute and solute-surface interactions determine the adsorption efficiency. Mortland et al. [33] compared these interactions in the adsorption of phenol and chlorophenols. They reported that the attraction of phenol to the hydrophobic surfaces of hexadecyltrimethylammonium- and hexadecyl-pridinium semectite was not sufficient to create a significant adsorption in the existence of water as a solvent. However, the adsorption on the same surfaces was considerable when a less polar solvent was used. This behavior was attributed to the less energetic interaction of phenol with the solvent. On the other hand, the effect of the polar aromatic compounds on the arrangement of the alkyl chains in the interlayers is of importance since the arrangement of cations greatly influences the interlayer spacing and therefore adsorption properties. Sheng et al. [37, 38] studied the adsorption of polar aromatic solutes on the hexadecyltri-methylammonium exchanged clays. They concluded that the polar solutes solvated the cationic centers in such a way caused the interlayer expansion due to a more vertical orientation of the alkylammonium chains. The re-orientation of the alkylchains and the interlayer expansion are accompanied by adsorption on the vacated mineral surfaces and thus the multiple mechanisms control the sorption of NOCs from the aqueous solutions on organoclays [38]. Sheng and Boyd [39] further examined the effect of solute polarity in a most recent work. The isomers of dichlorobenzene that display varying polarities depending on the position of the chlorine atoms on the aromatic ring were used. It was observed that the sorption was enhanced by greater interlayer separation and higher solute polarity and the existence of a polar component as a cosolute has the same effect with the interlayer separation and solute polarity.

Another important parameter is the sieving effect; the size and shape of the organic cation may exclude the adsorbate molecules [33]. Lawrence et al. [40] studied the relative organophilicity of tetramethylphosphonium and tetramethylammonium-smectite and the molecular size exclusion phenomena. They concluded that neither interlayer spacing nor surface area is responsible for the difference in adsorption behavior of phenols on the two organoclays and also the sorption is not strongly dependent upon the water solubilities of the adsorbates. Therefore, the steric exclusion should be considered, especially the placement of chlorine atoms in the orto, and meta position will cause the steric exclusion due to the increase in the width of the molecule.

Charge density and residence exchange cation of clay are other parameters affecting the extent of cation exchange and therefore adsorption of phenolics from the solution. The surface coverage depends on the charge density. For example, the stoichiometric adsorption of benzyl trimethylammonium occurs until all exchange sites on low charge and only 75% of those on high charge smectite are satisfied [41]. In the case of hexadeyltrimethylammonium, adsorptions occur beyond the cation exchange capacity for low and high charge smectites [41]. The orientation of alkylammonium cations on the surface will affect the adsorption by changing the mechanism and creating the sieving effect. Thus the surface charge of the clay should be considered in the preparation of organoclay for certain application.

Since wastewater can contain a mixture of phenolic compounds, the competition between the solutes is another parameter that should be considered. This competition results from the lower solubilities of phenolic compounds than phenol and their different pKa values causing the ionization at different pH values. Kim et al. [42] studied the binary and trisolute sorption systems containing 2-chlorophenol, 3-cyanophenol and 4-nitrophenol. They found that the sorption amounts of the solutes were less than that of each solute in a single solute system. The greater reduction in the sorption amount was observed in the adsorption of solute with a lower sorption affinity in the single solute system.

The surface structure, the interactions between the organic layer on the adsorbent and adsorbate, the modification procedures, time pH, temperature, and initial adsorbate concentrations are widely investigated in the adsorption studies. From an industrial view point, the amount of adsorbent used is of importance but there are only few papers reporting the effect of the amount of adsorbent [36, 43-47]. Lin and Cheng [36] reported that the optimum ratio of the amount of organobentonite to that of phenol or chlorophenol was 8 g/g. Shen [44] and Al-Asheh et al. [45] reported that the increase in the amount of removed phenol with increasing adsorbent concentration was limited. Yapar et al. [46] studied the effect of the amount of adsorbent and a nonlinear relation was observed between the adsorbent concentration and percentage removed. The nonlinearity was attributed to the formation of an additional resistance at high adsorbent concentrations. Due to this, the amount of adsorbents needs to be optimized for practical applications.

The discharge of pollutants also causes the contamination of soils since the pollutants are spread by surface and ground waters or as they directly come into contact with the soil. Therefore, the remediation of contaminated soils is another issue in environmental protection. This issue is handled by *in situ* applications or by preventing the spread of pollutants. The use of reactive barriers, which are based on a combination of adsorption and biodegradation, is a beneficial solution for the prevention of spreading by providing the elimination of target contaminants in the barriers. Witthuhn et al. [47] investigated the use of organoclays as reclaimable adsorbents for 2,4-dichlorophenol, when combined with biodegradation. Dodecyltrimethylammonium (C_{18}) and dioctadecyldimethylammonium ($2C_{18}$) were used in the preparation of organoclays to have different adsorbed surfactant layer characteristics. They concluded that the investigation gave encouraging results for the combination of organoclays with biodegradation and $2C_{18}$-montmorillonite could be used safely since the level of dissolved 2,4-dichlorophenol is well below the toxic level at the end of adsorption and intercalation.

The preparation of a suitable adsorbent first and then the use of it in adsorption studies is the general attitude in the removal of pollutants. In contrary of this attitude Shen [40] proposed a simultaneous process. The coagulation and adsorption properties of cationic surfactant modified bentonites are combined with the swelling of crude bentonite in this method. The process involves three steps: dispersion of bentonite in the contaminated water, the addition of a short chain alkylammonium cation and sedimentation. Dispersion of bentonite in water results in swelling and delamination and therefore a large surface area is exhibited. Beside the modification, the addition of a short chain cationic surfactant causes charge neutralization of bentonite and subsequent coagulation. The flocs, which are an agglomerate of organobentonite and bound NOCs, are removed by sedimentation. It was concluded that 90% phenol and nearly 100 % bentonite recovery could be achieved using benzyltrimethylammonium-bentonite under appropriate operating conditions and bentonite can be used as a recyclable support for the surfactant after the combustion of organic pollutants.

The concentrations of pollutants in the effluents of various industries are too low to be recovered economically and too high to discharge to the environment. Therefore, they have to be removed before discharging. Adsorption is a tool for the removal of the pollutants as they are accumulated on the adsorbent surface at relatively high concentrations. As a result, adsorption is actually the transfer of pollutants from one medium to another and thus a new round of environmental problems is created. This is the dilemma of adsorption and it can be corrected by the disposal of the pollutant-laden adsorbents. The solvent or thermal regeneration processes can be used in the disposal of adsorbents and the feasibility of the adsorption process depends on the cost of regeneration. In the case of solvent regeneration, the interactions among the solvent, adsorbed pollutants and organic layer on the adsorbent surface are of importance. First of all, the solvent should not cause desorption of alkylammonium cation on the surface and the adsorbed pollutants should have higher affinity to the solvent than the organic layer on the adsorbent surface. Boyd et al. [48] studied the desorption of pentachlorophenol and 3, 4, 5-trichlorophenol from hexadecyltrimethyl-ammonium- and dioctadecyldimethylammonium-smectites and carbon. The adsorbed species were exctracted with three different solvents using a Soxhlet extractor. It was observed that the efficiency of extraction varied with solvent for the extraction of chlorophenols from the organoclays used, than from carbon. Thermal regeneration is carried out in two ways: by combustion of the pollutants and/or by removing them at relatively low temperatures. Beside the renovation of modification after each regeneration cycle, combustion will cause changes in some structural properties of the clay. For example, the inner crystalline swelling and rehydration properties will be negatively affected by heating at temperatures higher than 300°C [21]. The change in the rehydration characteristics will create difficulties in the renovation of modification and probably will cause to a lesser extent of cation exchange resulting with the different surface structure and a decrease in the adsorption efficiency relative to the unused adsorbent. Thus, the use of regenerated adsorbent will not be feasible after a few regeneration cycles and the change with unused adsorbent will be required. In case the small numbers of regeneration cycles, the use of clay as a recyclable support will not be efficient. The thermal regeneration that is carried out at low temperatures will be considered as a solution to this problem. Lin and Cheng [36] studied the low temperature thermal regeneration of phenol and m-chlorophenol laden organobentonites. The regeneration tests were conducted in a nitrogen atmosphere to minimize the oxidative decomposition of quaternary amines on the surface. It was concluded that 200°C and 2h were good operating conditions and the observed color change at temperatures higher than 200°C was attributed to phenol and chlorophenol decomposition. In another work, Lin and Cheng [49] also reported that the energy supplied at temperatures lower than 200°C was not sufficient to remove all phenol from the organobentonite surfaces by breaking all the physical bonds and excessive temperatures caused structural changes in organobentonite leading to a lower phenol adsorption in the regenerated organobentonites.

As a result, it is possible to prepare an organoclay complex with desired characteristics by taking into account the properties of the phenolic compounds to be removed. However, the feasibility of the adsorbents depends on the effectiveness of the regeneration process in some respects, e.g the cost of regeneration, the number of regeneration cycles etc. and the current regeneration processes of organo bentonites do not seem ultimate solution. So, either new regeneration process or the new types of adsorbents with higher thermal stabilities should be developed.

2.2. Organo-Inorgano Modified Clays

Pillared clays, PILCs, are obtained by introducing polyhydroxo metal cations through the ion exchange. In addition to their unique polarity, pore size distribution and high surface areas; pillared clays have high thermal stability giving them superiority in regeneration. As in the case of organoclays, the exchange sites are occupied in the pillared clays; they are also prepared through the replacement of neutralizing cations. In spite of all these similarities in the way of preparation, the resulting structures have quite different mechanical properties and the pillar, is the very word stressing these differences, since it means column for supporting part of a structure. Pillars provide a permanent separation between the layers of the clay and a zeolite like structure is formed [50, 51]. The pillaring reaction actually includes two steps: 1) preparation of a bulky polyoxyhydroxy cation such as $[Al_{13}O_4(OH)_{24}(H_2O)_{12}]^{7+}$ the so called Keggin ion and 2)the intercalation of the cation in the host mineral, clay. Polyoxyhydroxy cation is a product of hydrolysis reaction, which can be carried out in the solution or interlamellar space of the clay [49]. Different methods can be used to perform the hydrolysis in the solution. They are: 1) addition of metal carbonates (Na, Mg, Zn, etc.) to a solution of

$AlCl_3$ or $FeCl_3$, 2) addition of metal hydroxides, 3) addition of metallic aluminium in HCl and/or $AlCl_3$, and 4) electrolysis of a 1M solution of $AlCl_3$ [51]. Through the mixing of clay and polycation, the intercalation is carried out and the interlamellar space is segmented to form additional pores. The size of the pores depends on the amount of polycation, which is determined by the method of preparation, nature of the reactants, their initial concentration, the degree of hydrolysis (the OH/M^{n+} ratio), the reaction temperature, the rate of the addition of reactants and the aging conditions of the hydrolized solutions [51]. Beside the formation of extra pores, the basal spacing is also increased and the resulting material has large surface area and pore volumes, being potentially suitable for applications as catalysts and adsorbents [52]. Although Al, Fe and Zr are used as pillaring agents, there is more information on properties of Al-pillared clays since their utilizations in catalytic applications and adsorption are more common. Zielke and Pinnavaia [53] prepared alumina-pillared, chromia-pillared montmorillonites and alumina delaminated laponite to determine the adsorption properties for the adsorption of chlorinated phenols. They observed that binding of neutral pentachlorophenol at pH 4.7 to all adsorbents never reached saturation values. Differences in the pentachlorophenol binding capacities for three adsorbents attributed to the dependence of adsorption on the pore structure and surface composition of the adsorbent in addition to the surface area. Danis et al. [54] carried out a similar work by using the different substituents of chlorinated phenols and the adsorption properties of Al-pillared montmorillonite, which is a microporous material, were compared to those of mesoporous alumina aluminum phosphates. They also reported that the adsorption was dependent on the pore structure and surface composition of the adsorbent. For similar specific surface areas, it was observed that Al-pillared montmorillonite has a higher adsorption capacity than the phosphates in spite of their large pores. As in the case of adsorption, desorption from alumina aluminum phosphates was also lower as compared with Al-pillared montmorillonite. The desorption of 2,4,6-trichlorophenol and pentachlorophenol from Al-pillared montmorillonite by acetone was over 73.9% with regard to the adsorbed amount. It was concluded that the surface chemical effects might be very useful in the designing a variable adsorbent for ground water treatment applications.

As already explained, the existence of carbonaceous or hydrophobic surfaces has a significant effect on adsorption of NOCs by various adsorbents. Thus the addition of the organic compounds such as cationic and/or nonionic surfactants will be beneficial in the improvement of adsorption properties of PILCs. Cationic surfactants have pronounced affinity to the solid surfaces and they are widely used in the preparation of organoclays. Although the nonionic surfactants do not have any charge, they tend to accumulate on the solid surfaces via hydrophobic interactions. Srinivasan and Fogler [55, 56] reported the results of their researches on the preparation and use of inorgano-organo-clays (IOCs) in the removal of priority pollutants from industrial wastewaters. IOCs were prepared by blocking cation exchange sites of montmorillonite with inorganic ions and the succesive adsorption of cationic surfactants. In the first part of the research, adsorption kinetics and adsorption-desorption behaviors of surfactants, nonexchangeability of polycations, flocculation-peptization of IOCs and sequential desorption of adsorbed surfactants were investigated. At the end of the experiments, it was decided that about 90% of surfactants were irreversibly bound to the surface and the adsorbed surfactant moiety was oriented with its hydrocarbon tails towards the surface. In the second part, the adsorption effeciencies of the IOCs were compared with granulated activated carbon and cetylpridinium cation exchanged montmorillonite. It was concluded that the IOCs were more efficient than the tradional cation exchanged smectite and they can be thought of as viable cost effective alternatives of activated carbon in wastewater treatment applications.

Surface area and cation exchange capacity are properties specific to the type of clay used. Therefore any change in the type of mineral and modification conditions will yield a different structure and consequently different adsorption behavior. Both cation exchange capacity and the amount of polycations determine the surface configuration and thus the pore size distribution. It is possible to change the amount of polycations by using different OH/M^{n+} ratios in the preparation of the cation. If the surfactant is added after the preparation of the PILCs, the surfactant molecules will be adsorbed on the exposed surface between the pillars, on the pillars and finally on the external surfaces. If the surfactant is added during the preparation, it is possibly adsorbed on the cation exchange sites, on the surface of the pillars and on the external surface. The distribution of surfactant between all these phases depends on the amount of surfactant and modification conditions. Therefore, the metal: surfactant: clay ratio should also be considered in the preparation of IOCs, since the adsorption behavior will change significantly. In the case of phenol adsorption, the adsorption capacity will increase with increasing surfactant clay ratio because the adsorption occurs mainly in the carbonaceous layer on the surface. Except the effect of surface area, this has been confirmed in a study conducted by Jiang and Zeng [57]. They observed that the montmorillonite KSF, which has a surface area around 20-40 m^2/g performed better than another ones with a greater surface area. This behavior was attributed to the differences in the crystalline phase. It was also observed that there is an optimum in metal:surfactant:clay ratio and it was determined to be 4:0.2:1 for the adsorption of phenol on IOCs obtained by modification of Al/Fe pillared clay with hexadecyltrimethylammonium.

Although the nonionic surfactants are also adsorbed on solid surfaces via hydrophobic interactions, the resulting surface structure will be completely different. First of all, the lack of net charge will cause a different configuration and the filling of micropores completely with surfactant molecules at high loadings. The filling of micropores with surfactant molecules will create the differences in the adsorption behavior. Michot and Pinnavia [58] studied the effect of surfactant loading on the adsorption of phenol on Tergitol modified pillared clays. They observed that the presence of surfactants in the micropores of the pillared clay did not enhance the adsorbed amount of 3-chlorophenol and the surfactant- modified external surfaces have a high affinity for 3-chlorophenol. In the case of cationic surfactants, it is not possible to fill the micropores completely due to the repulsion between the positive charges. However the similarities such as the changed basal spacing also exist. In general an increase in the basal spacing is observed regardless of the type of modification but the magnitude of this increase is closely related to the swelling behavior of the clay [57]. For a certain type of clay, the basal spacing increases with the surfactant loading in both surfactant-modified and inorganic organic clays. In contrary to the increase in basal spacing, the surface area decreases with the surfactant loading for both types of modified clays; the clogging of the pores by alkylchains to the N_2 probe molecules is the plausible reason for this decrease.

The presence of phenolate, as a result of the dissociation of phenol at high pH values, will result in considerably less adsorption of phenol on the surfactant modified pillared clay due to the decrease in affinity of adsorbent towards the phenolate ions. This behavior was observed by Bouras et al. [59] in the removal of pentachlorophenol from aqueous solution by cetyltrimethylammonium chloride modified Fe- pillared bentonite and by Michot and

Pinnavaia [58] for the adsorption of chlorinated phenols on tergitol modified pillared clays. However, contrary observations were also reported [45, 60]. Wu et al. [60] observed two inflexions in the curves showing the effect of pH on adsorbing phenol by different montmorillonites, including crude and modified forms, and this behavior was explained by the presence of three different adsorption stages: physical, cation exchange and chemical. The physical adsorption stage covers the pH range from 3 to 8 at which phenol is in a neutral molecular form and it is adsorbed on the outer surface. Cation exchange adsorption occurs between pH values 8-10. Phenol is adsorbed on both outer and interlayer spaces and the adsorption increases. For the pH values higher than 10, the stage is chemical adsorption at which physical, cation exchange and surface acidity sites adsorptions occur. As a result, phenol removals increase with increasing pH.

In addition to the structural similarities to zeolites, high BET surface area, microporosity and Brönsted/Lewis acidity make the pillared clays the potential catalyst or catalyst precursors. Catalyst deactivation and regeneration are the major problems in catalytic reactions. Thermal treatment is widely used in the regeneration of catalysts; therefore thermal stability is a key property for the use of any porous material in catalytic applications. For these reasons the thermal stabilities of pillared clays are deeply investigated besides their catalytic activities. As reported by many researchers, the thermal treatment up to 800°C causes no considerable change in the basal spacing and BET surface area [52, 61, 62]. These results are good indications for the regeneration of exhausted PILCs and IOCs. The combustion of organic adsorbate at 500°C is the usual way followed in the regeneration process. In the case of surfactant modified pillared clays, the removed surfactants during the thermal treatment are reloaded before the next adsorption cycle. The efficiency and structure of the reloaded surfactant layer depend on the regeneration. Combustion can cause some changes in the surfactant adsorption capability of the surface and the structure on the surface. As a result of this, the efficiency of the adsorbent will possibly be different from the unused adsorbent. The dehydroxylation of the pillars causes slightly more surfactant adsorption and thus more removals from aqueous solutions [58].

3. Modeling of Adsorption Behavior

The research on adsorption is conducted in three parts: the preparation and characterization of the adsorbent, the adsorption kinetics and isotherm studies. Various techniques are used in the preparation and characterization of adsorbents. The method used in the preparation determines the ultimate adsorption properties. Pore size distribution, surface area, surface charge, interlayer distance and cation exchange capacity are the characteristics of an adsorbent. Pore size distribution and surface area are determined by nitrogen adsorption and interlayer distance by XRD analysis. Cation exchange capacity is determined by analytical procedures. The characterization of adsorbents gives very useful tips for the estimation of adsorption capability of a solid. By determining of the time required to reach the equilibrium and the change in adsorbed amount with the equilibrium concentration can be concluded whether the use of adsorbent in the removal of targeted pollutant will be feasible or not. The last step is the modeling of adsorption behavior by applying the kinetic and isotherm data to the relevant models. This step is required for elucidating the adsorption mechanism and also for the future design of a large-scale adsorption column.

3.1. Kinetic Studies

The determination of adsorption kinetics is of prime importance in the estimation of the rate at which the pollutant is removed from an aqueous solution. The rate of removal is an important parameter in order to design appropriate sorption treatment plants. Adsorption from a solution on porous adsorbents is a three-step process involving film diffusion, particle diffusion and adsorption on the interior surface of the adsorbent through mass action controlled mechanism [63]. According to Crank [64] adsorption on the interior surface is very rapid and it is not the rate determining step, therefore either film diffusion or particle diffusion will be the rate- determining step depending on the mixing, solute concentration and particle size of the adsorbent. Film diffusion is the rate-determining step for a system having a) poor mixing, b) dilute solute concentration, c) small particle size of adsorbent and, d) high affinity of solute for adsorbent. Otherwise, particle diffusion is the rate-controlling step. The kinetic data can also be used to determine which step controls the rate. The particle diffusion will be the rate-determining step, when a plot of adsorbed amount versus the square root of time is linear and it passes through the origin [65]. If the results are represented by a linear relationship but the line does not pass the origin, both the particle diffusion and the other mechanisms are the rate determining steps [66].

The equations used in the modeling are either diffusion based or rate law based. In the case of diffusion-based models, a two-step general resistance model is described first [67] and then the equation is modified according to the conditions. The rate law based equations are determined by experimentation and they cannot be inferred by more examination of the overall chemical reaction equation [68]. Among the model based rate equations, the pseudo first order rate expression of Lagergren [69] is widely used. It has the form,

$$\frac{dq_t}{dt} = ks_1(q_1 - q_t) \qquad (1)$$

Integration for the boundary conditions t=0 to t=t and q_t=0 to q_t= q_t gives:

$$\log(q_1 - q_t) = \log(q_1) - \frac{ks_1}{2.303}t \qquad (2)$$

where,ks_1 is the rate constant of first order sorption (1/min), q_1 is the adsorbed amount at the equilibrium (mg/g), q_t is the adsorbed amount at any time t, (mg/g).

If the adsorption involves rather complex interactions and the effects of transport phenomena and chemical reactions are not experimentally inseperable [70], the pseudo second order model is used. The model is based on the assumption that the sorption capacity is proportional to the number of the active sites occupied on the sorbent [68]. The proposed model has the form,

$$\frac{dq_t}{dt} = k(q_c - q_t)^2 \tag{3}$$

Separating the variables and integrating for the boundary conditions t=0 to t=t and q_t=0 to q_t= q_t gives,

$$\frac{1}{(q_c - q_t)} = \frac{1}{q_c} + kt \tag{4}$$

Where k is the rate constant of sorption (g/mg min), q_c is the adsorbed amount at the equilibrium (mg/g), q_t is the adsorbed amount at any time t, (mg/g).

There is not much work on the kinetics of phenol adsorption in literature. Although the change in the adsorbed amount with time is studied in the experimental studies, the determination of time required to reach the equilibrium is the main objective. Banat et al. [71] studied the effect of time in the adsorption of phenol on bentonite. They reported that the time required for the equilibrium to be established is 6h, particle diffusion is involved in the sorption process with some other mechanisms but no rate constant is reported. In another work, the time required for the equilibrium on the adsorption of methylphenols on a carbonaceous adsorbent is reported as 8 h and the application of Lagergren's equation is possible [13]. It was also found that the adsorption process is a first order process. Juang et al. [32] applied the pseudo-second order model to the adsorption of phenol, m-nitrophenol and o-cresol on cetyltrimethylammonium-montmorillonite. They concluded that the model describes the sorption process. Yılmaz and Yapar[34] studied the adsorption kinetics of phenol on tetradecyltrimethylammonium and hexadecyltrimethylammonium bentonites with the exchange degrees of 25%, 50% and 100%. It was reported that the adsorption of phenol on each adsorbent is described by a pseudo second order equation. Kinetic studies are generally carried out by using a batch method; by taking the samples at certain time intervals. In contrary of this usual way Zhang and Sparks [71] carried out the kinetic studies using a stirred-flow reaction chamber. They studied both adsorption and desorption kinetics by applying an unsteady-state mass transfer equation that the cumulative amount adsorbed is expressed in terms of the differences between the accumulation of the adsorbate and the change in the average adsorbate concentration in the reaction chamber at the time interval t+Δt. It was observed that the considerable amount of desorption; about 42% of adsorbed phenol occured in 30 min in the case of desorption from hexadecyltrimethylammonium-montmorillonite.

3.2. Isotherm Studies

The industrial scale adsorption processes are carried out in batch, semicontinous or continuous adsorbers. The design of such a large-scale adsorber requires information on how the adsorbed amount changes with the equilibrium bulk concentration as well as the rate of adsorption. An adsorption isotherm is actually a mathematical model giving this information. Since the adsorption systems are unique, the type and constants of the isotherm equation should be determined by conducting experimental work. The work involves the determination of equilibrium solute concentration and the corresponding adsorbed amounts by using batch or continuous processes. The isotherm model should represent this data within experimental uncertainty, in other words, the correlation coefficient should be high. Beside the high correlation coefficient, the equation must satisfy a necessary thermodynamic boundary condition, Henry's law: at very low surface coverage the equation must show that the equilibrium adsorbed amount is proportional to the equilibrium concentration [72]. In terms of design considerations, the proposed model should be as simple as possible and contain less parameter.

The Langmuir and Freundlich isotherms are the most frequently used in the mathematical representation of the data obtained in the removal of phenolic compounds by using crude-, organo- and inorgano-organo clays. In addition to these two parameter nonlinear equations, the use of a three -parameter Redlich-Peterson isotherm is also common.

The Langmuir isotherm is expressed by,

$$\theta = \frac{Q_e}{Q_{mon}} = \frac{K_L C_e}{(1 + K_L C_e)} \tag{5}$$

where θ, Q_e(mg/g), Q_{mon}(mg/g), C_e (mg/L), K_L(L/mg) represent the fractional coverage, amount adsorbed at equilibrium, amount adsorbed corresponding to monolayer coverage, equilibrium bulk concentration and Langmuir constant, respectively. The linear form of the Langmuir isotherm is represented by

$$\theta = \frac{C_e}{Q_e} = \frac{C_e}{Q_{mon}} + \frac{1}{K_L Q_{mon}} \tag{6}$$

or

$$\frac{1}{Q_e} = \frac{1}{K_L Q_{mon} C_e} + \frac{1}{Q_{mon}} \tag{7}$$

The Freundlich model is an empirical isotherm model and it is suitable for highly heterogeneous surfaces [73]. It does not approach Henry's law and provides a poor fit for the adsorption data at low concentrations [72]. However it gives a good fit for the adsorption data at moderate concentrations. The equation for the Freundlich isotherm is

$$Q_e = K_F C_e^{1/n} \tag{8}$$

where K_F(mg^{n+1}, g^{-1},L^{-n}) and 1/n are Freundlich constants related to adsorption capacity and adsorption intensity of the adsorbent, respectively. The linear form is represented as,

$$\ln Q_e = \ln K_F + \frac{1}{n}\ln C_e \tag{9}$$

The Redlich-Peterson equation incorporates the features of the Langmuir and Freundlich isotherms. It can reduce to Henry's law at low concentrations and approaches Freundlich isotherm at high concentrations [72]. The isotherm is described as

$$Q_e = \frac{a_R C_e}{1 + b_R C_e^{\beta}} \tag{10}$$

where a_R (L/g), b_R, $(L\ mg^{-1})^{\beta}$ and β ($0<\beta<1$) are constants.

In addition to above mentioned isotherm models, the use of dual-sorption and a three-parameter empirical isotherm models are also reported [36, 74]. The dual sorption model has the form,

$$Q_e = K_p C_e + \frac{Q_m K_L C_e}{1 + K_L C_e} \tag{11}$$

and K_p, Q_m, K_L are the parameters.

The three-parameter empirical model based on the phenomenological description of sorption data in organoclay approaches Henry's law and the Freundlich isotherm model in the low and high concentrations, respectively. The equation for the model is,

$$Q_e = K\left(1 + \beta C_e^2\right)^{(n-1)/2} C_e \tag{12}$$

where K, β and n are the three parameters involved in the model.

The nonlinear and linear forms of all above-mentioned expressions are used in the mathematical presentation of isotherm data. Viraraghavan and Alfaro [8] used both forms in the adsorption of phenol on bentonite. The correlation coefficients for linear and the nonlinear forms of the isotherm were found to be 0.86 and 0.80, respectively. This result represents relatively large deviations from the experimental values or a relatively poor fitting. Banat et al.[64] reported the correlation coefficients as 0.996, 0.904 and 0.948 for the linear form of the Freundlich isotherm in the case of phenol adsorption on bentonite at three different pH values. A comparison of the results of the two works reveals that the experimental data are represented by the same isotherm with different isotherm constants and correlation coefficients. This result points to similar adsorption mechanisms but relatively different adsorbate-adsorbent and adsorbate-solution interactions. In the case of adsorption of p-chlorophenol on bentonite, it was determined that the Langmuir isotherm fits better than the Freundlich isotherm to the experimental data [23]. This result confirms the effect of adsorbate-adsorbent and adsorbate-solvent interactions.

Lin and Cheng [36] reported adsorption of phenol and m-chlorophenol on the organobentonites prepared by using short- and long chain alkylammonium cations and one containing an aromatic hydrocarbon group, which was described by the Freundlich equation with relatively high correlation coefficients. The highest K_F values are obtained for the adsorption of phenol and chlorophenol on the benzyltrimethylammonium-bentonite and they are close to each other. This result indicates that the structural similarities between the pseudo organic phase on the surface and adsorbate cause stronger interactions between surface and adsorbate. Yapar et al.[46] observed that the adsorbed amount changes significantly with small changes in equilibrium bulk concentration and this behavior is reflected in the Freundlich constant 1/n, which is a measure of the deviation from the linearity of adsorption.

The major constituent of bentonite is the clay mineral montmorillonite, therefore the organo-bentonites and montmorillonites are approximately the same complexes. The differences in cation exchange capacities and other surface properties create the changes in the orientation of the organic cation on the surfaces; however, this is also the case with organobentonites. Juang et al.[32] used Langmuir, Redlich-Peterson and dual- mode sorption models in the mathematical representation of single solid sorption onto cetyltrimethylammonium-montmorillonite. For the adsorption of phenol, m-nitrophenol and o-cresol it was found that the correlation coefficients were the highest in the case of the dual-mode sorption model. Kim et al [42] reported the use of Langmuir and Redlich-Peterson models to fit the single- solute sorption equilibria and ideal adsorbed solution theory, IAST, coupled with a single solute model. It was decided that the predictions from IAST yielded favorable representations of the multisolute competitive sorption of phenolic compounds onto hexadecyltrimethylammonium-montmorillonite. Song and Shin [74] proposed a new three - parameter model and they compared the goodness of fit of the new model with four nonlinear models including Langmuir, Freundlich, Redlich-Peterson and dual-mode. It was decided that the three-parameter models were better than the two-parameter models in describing the data for the adsorption of 2-chloro-, 3-cyano-, and 4-nitrophenol on organo-montmorillonites. However no preference to the Song model over Redlich-Peterson and dual-mode sorption models in terms of goodness-of-fit was mentioned.

Srinivasan and Fogler [56] analyzed the adsorption of pentachlorophenol on cetylpyridinium- Al-, and cetylpyridinium- La-montmorillonites by using the linearized form of the Langmuir isotherm. It was observed that the data points at larger values of equilibrium concentrations do not conform to the Langmuir isotherm. This deviation was attributed to the existence of surface sites having multiplicity of adsorption free energies. Wu et al. [60] reported that the Freundlich isotherm could be used in the modeling of the adsorption behavior of phenol on crude-, pillared-, organo-inorgano-montmorillonite. The values of K_F were found to be, 3.936; 2.244; 3.446 and 5.936 for crude-, pillared-, organo-inorgano- and organomontmorillonite, respectively. According to this result, the organomontmorillonite has the highest adsorption capacity. The goodness of fit of the Freundlich equation was also reported for the adsorption of phenol on hexadecyltrimethylammonium montmorillonite and IOCs [51]. A summary of the isotherm models used in the mathematical representation in the adsorption data of phenolics and the correlation coefficients are given in the Table 3.

As shown in the Table, it is not possible to represent the isotherm data by one isotherm model even for the same sort of adsorbent and adsorbate. It is also not possible to classify the adsorption process depending on the type of isotherm. Every system is actually unique and this property requires the determination of the adsorption isotherm and its parameters for the system considered for industrial applications. The design of a single stage process requires the knowledge on how the concentration changes at the inlet and the exit of the stage and what is the corresponding equilibrium adsorbed amount. A single stage adsorption process is not sufficient to reach the desired separation and the stagewise material balance equations are more complex in multistage applications since the total adsorbed amount is expressed as the The design of a single stage process requires the sum of adsorbed amount at the exit of each stage. For example for a two stage process [75],

$$q_{e1} + q_{ef} = L[(C_0 - C_{e1})/M_1 + (C_{e1} - C_{ef})/M_2] \tag{13}$$

where, q_{e1} and q_{ef} are the organic concentrations (mg/g) in the solid (adsorbent) phase at the exit of first and second stages.C_0 is the organic concentration (mg/L) in the feed.M_1 and M_2 are the amount (g) of adsorbent C_{e1} and C_{ef} are the organic concentrations (mg/L) at the exit of the first and second stages.

Table 3. The Isotherm Models and Their Correlation Coefficients

Adsorbent	Adsorbate	Isotherm	Correlation Coefficient, R^2	Ref. No
Bentonite	Phenol	Freundlich Freundlich nonlinear	0.86 0.80	8
Bentonite	Phenol	Freundlich Langmuir	0.904-0.996 0941-0.985	65
Bentonite	p-chlorophenol	Freundlich Langmuir	0.987 0.997	23
Organo Bentonites	Phenol m-chlorophenol	Freundlich Freundlich	0.978-0.994 0.989-0.998	36
Organo Bentonite	Phenol	Freundlich	0.989	46
Organo Montmorillonite	Phenol m-nitrophenol	Langmuir Redlich-Peterson Dual Mode Sorption Langmuir Redlich-Peterson Dual Mode Sorption	0.9958 0.9933 0.9972 0.9852 0.9861 0.9922	32
Organo Montmorillonite	2-chlorophenol 3-cyanophenol 4-nitrophenol	Langmuir Redlich-Peterson Langmuir Redlich-Peterson Langmuir Redlich-Peterson	0.9961-0.9989 0.9987-0.9989 0.9758-0.9974 0.9969-0.9982 0.9896-0.9956 0.9988-0.9997	42
Organo Montmorillonite	2-chlorophenol 3-cyanophenol 4-nitrophenol	Song Model	0.9940-0.9995 0.9980-0.9988 0.9954-0.9997	74
Organo-inorgano Montmorillonite	Pentachlorophenol 3,5-dichlorophenol	Langmuir Langmuir	0.99 0.99	56
Organo-inorgano Montmorillonite	Phenol	Freundlich	Not mentioned	60
Organo montmorillonite Pillared montmorillonite Organo-inorgano Montmorillonite	Phenol	Freundlich	0.94 0.94 0.96-0.98	51

Equation 13 is the combination of the material balances around the stages, in other words it is the combination of operating lines. The compositions of the streams leaving a stage are found by using the operating lines and the equilibrium curve. For multistage operations the graphical representation of the stages will be difficult and the relation between the equilibrium adsorbed amount and equilibrium concentration is inserted into the material balance. In the design of an adsorber the Freundlich model is frequently applied [76] but the model does not describe the adsorption isotherm satisfactorily as mentioned previously. The use of other isotherm models, such as Langmuir and Redlich-Peterson can be satisfactory but it may cause an increase in the complexity of the material balance equations. Therefore, the simplicity and the consistency of the isotherm model offered are of importance. However, these two constraints are satisfied for a few systems at the same time; either the equation is simple and the correlation coefficient is low or the correlation coefficient is high and equation is not simple. In this respect, the first step should be to make a decision on what should be the minimum acceptable value of the correlation coefficient or the complexity of the isotherm equation.

Lin et al. [75] studied the counter current multistage phenol adsorption on organo bentonites synthesized using hexadecyltrimethylammonium, benzyltriehylammonium and cetypridinuim bromides. The experimental tests were conducted by using 2 grams of adsorbents in single-, two- and three- stage adsorption processes. It was reported that the agreement between the calculated and the observed phenol removal efficiencies are quite good for single, two and three stage processes. The difference between the theoretical predictions and the measured data is less than 2% for single and two stage processes. A significant improvement was also observed in the phenol removal efficiency for both two- and three-stage processes in comparison with the single stage. As a matter of fact, the amount of adsorbent required to reach a predetermined removal rates decreases with the increasing number of stages in a cascade at first but approaches the minimum value only asymptotically [76]. However, the use of more than two stages in a counter current cascade is not economical since intermediate filtration is required between the stages in industrial applications [76]. In addition to the labor, the storage of the pollutant-laden adsorbent and the possible deterioration of the adsorbent and/or laden pollutants during storage are the factors that increase the operating cost. Therefore, the use of a crosscurrent operation is more practical. The other problems encountered in a batch process are arising from the resistances to the mass transfer. As mentioned previously there are three different resistances; film diffusion, particle diffusion and adsorption on the interior surface of the adsorbent through mass action controlled mechanisms. Since the film diffusion resistance is eliminated under sufficient mixing conditions, the determination of the agitation is important. This can be done by conducting both small- and large –scale tests.

4. CONLUSION

It has been concluded that;

- the adsorption of phenolic compounds on mineral surfaces in their neutral molecular forms is principally governed by solute-solvent and solute-surface interactions.

- The pore size has an important effect on the adsorbed amount in the adsorption on a clinoptilolite surface.
- The hydrophobicity of the surface is another parameter determining the adsorption capability of a potential adsorbent to be used for the removal of phenolic compounds.
- The existence of a carbonaceous layer plays an important role in the adsorption.
- The adsorption mechanism depends on the structure of the pseudo organic layer on the surface of organoclays. In the case of short chain alkylammonium cations, the adsorption of phenolic compounds occurs by hydrogen bonds to the water molecules of the inorganic interlayer cation and van der Waals attraction to alkyl chains. In the case of long chain cations, the mechanism of adsorption is partitioning.
- The amount of adsorbents needs to be optimized for practical applications.
- It is possible to prepare an organoclay having the predetermined properties which are dictated by the physicochemical properties of the phenolic compounds to be removed.
- Regeneration of organoclays is an important issue due to the changes in clay structure occurring at high temperatures and causing the formation of a different surface structure during the renovation of the modification.
- Metal: surfactant: clay ratio should be considered in the preparation of IOCs since the adsorption behavior will change significantly depending on this ratio.
- It is not possible to represent isotherm data by one isotherm model using even the same kind of adsorbent and adsorbate.
- The model best fitted and its parameters are determined by conducting laboratory tests.
- The final conclusion on the subject is that the modified clays are innovative and promising adsorbents in the removal of phenolic compounds.

REFERENCES

[1] Ahmaruzzaman, M., Sharma, D.K., *Journal of Colloid and Interface Science*, 2005, 287, 14-24.

[2] Muller, F., Caillard, L., *In Ullman's Encyclopedia of Industrial Chemistry*, 5 th Edition, VCH, Weinheim, 1986,Vol A7.

[3] Beall, G.B., *Applied Clay Science,* 2003, 24, 11-20.

[4] Kujawski, W., Warszawski, A., Ratajczak, W., Porebski, T., Capala, W., Ostrowska, I., *Desalination*, 2004,163, 287-296.

[5] Al-Shafey, E.I., Correia, P.F.M., Carvalho, J.M.R., , *Seperation Science and Technology,* 2005, 40, 2841-2869.

[6] Tanthapanichakoon, W., Ariyadejawanich, P., Japthong, P., Nakagawa, K., Mukai, S.R., Tamon, H., *Water Research*, 2005, 39, 1347-1353.

[7] Mooney, D.A., Muller-Plathe, F., Kreneg, K., *Chem. Phys. Lett*., 294, 135, 1998.

[8] Viraraghavan, T., Alfaro, F.M.,*Journal of Hazardous Materials*, 1998, 57, 59-70.

[9] Kao, P-C., Tzeng, J-H., Huang, T-L., *J. of Hazardous Materials*, 2000,76, 237-249.

[10] Sarkar, M., Acharya, P.K., Bhattacharya, B., *Journal of Chemical Technology and Biotechnology*, 2005, 80, 1349-1355.

[11] Srivasta, V.C., Swamy, M.M., Mall, I.D., Prasad, B., Mishra, I.M., Colloids and Surfaces A: *Physicochem. Eng. Aspects*, 2006, 272, 89-104.
[12] Mott, H.V., Walter, J., Weber, Jr., Environ. Sci. Technol., 1992, 26, 1234-1242.
[13] Jain, A.K., Suhas, A., Bhatnagar, A., *Journal of Colloid Interface Science*, 2002, 251, 39-45.
[14] Sivaraj, R., Namasivayam, C., Kadirvelu, K., *Waste Management*, 2001, 21, 105-110.
[15] Ho, Y-S., *Water Research*, 2003, 37, 2323-2330.
[16] Wang, X-S., Qin, Y., *Process Biochemistry*, 2005, 40(2), 677-680.
[17] Aksu, Z., Kutsal, T., Gün, S., Hacıosmanoğlu, N., Gholaminejad, M., *Environ. Technol.*, 1991, 12, 915-921.
[18] Aksu, Z., Yener, J., *Process Biochemistry*, 1998, 33(6), 49-655.
[19] Rao, J.R., Viraraghavan, T., *Bioresource Technology*, 2002,85, 165-171.
[20] Wu, J., Yu, H-Q., *Process Biochemistry*, 2006, 41, 44-49.
[21] Grim, R.E., Clay Mineralogy, 1968, 78, Mc Graw-Hill, Inc. New York.
[22] Şişmanoğlu, T., Pura, S., Colloids and Surfaces A: *Physicochem. Eng. Aspects*, 2001, 180, 1-6.
[23] Kuomanova, B., Peeva-Antova, P., *Journal of Hazardous Materials*, 2002, A90, 229-234.
[24] Khalid, M., Joly, G., Renaud, A., Magnoux, P., *Ind. Eng. Chem. Res.*, 2004, 43, 5275-5280.
[25] Roostaei, N., Tezel, F.H., *Journal of Environmental Management*, 2004, 70, 157-164.
[26] Li, Z., Burt, T., Bowman, R.S., *Environ. Sci. Technol.*, 2000, 34, 37556-3760.
[27] Bergaya, F., Lagaly, G., *Applied Clay Science*, 2001, 19, 1-3.
[28] Lagaly, G., *Progr. Colloid and Polym. Sci.*, 1994, 95, 61-72.
[29] Lagaly, G., Weiss, A., Kolloid-Zeitschriftand Zeitschrift für polymere, 1970, 243, 48-55.
[30] Dekany,I., Szanto, F., Nagy, L.G., *Journal of Colloid Interface Science*, 1986,109(2), 376-384.
[31] Koh, S-M, Dixon, J.B., *Applied Clay Science*, 2001, 18, 111-122.
[32] Juang, R-S., Lin, S-H., Tsao, K-H., *Journal of Colloid and Interface Science*, 2002, 254, 234-241.
[33] Mortland, M.M., Shaobai, S., Boyd, S.A., *Clays and Clay Minerals*, 1986, 34(5), 581-585.
[34] Yılmaz, N., Yapar, S., *Applied Clay Science*, 2004, 27, 223-228.
[35] Zhu, L., Ren, X., Yu, S., *Environ. Sci. Technol.*, 1998, 32, 3374-3378.
[36] Lin, S.H., Cheng, M.J., *Waste Management*, 2002, 22, 595-603.
[37] Sheng, G., Xu, S., Boyd, S.A., *Water Research*, 1996, 30(6), 1483-1489.
[38] Sheng, G., Xu, S., Boyd, S.A., *Environ. Sci. Technol.*, 1996, 30, 1553-1557.
[39] Sheng, G., Boyd, S.A., *Clays and Clay Minerals*, 2000, 48(1), 43-50.
[40] Lawrence, M.A.M., Kukkadapu, R.K., Boyd, S.A., *Applied Clay Science*, 1998, 13, 13-20.
[41] Shen, Y-H., Colloids and Surfaces A: *Physicochem. Eng. Aspects*, 2004, 232, 143-149.
[42] Kim, D-G., Song, D-I., Jeon, Y-W., *Separation Science and Technology*, 2001, 36(14), 3159-3174.
[43] Al-Malah, K., Azzam, M.O.J., Abu-Lail, N.I., *Sep. Purif. Technol.*, 2000, 20, 225-234.
[44] Shen, Y-H., *Water Reseacrh*, 2002, 36, 1108-1114.

[45] Al-Asheh, S., Banat, F., Abu-Aitah, L., *Sep. Purif. Technol.*, 2003, 33, 1-10.
[46] Yapar, S., Özbudak, V., Dias, A., Lopes, A., *Journal of Hazardous Materials*, 2005, B121, 135-139.
[47] Witthuhn, B., Klauth, P., Klumpp, E., Narres, H-D., Martinius, H., *Applied Clay Science*, 2005, 28, 55-60.
[48] Boyd, S.A., Shaobai, S., Lee, J-F., Mortland, M.M., *Clays and Clay Minerals*, 1988, 36(2), 125-130.
[49] Lin, S.H., Cheng, M.J., *Environmental Technology*, 2000, 21, 475-482.
[50] de Andrés, A.M., Galvan, M.J.C., Ruiz-Hitzky, E., *Material Research Bulletin*, 1999, 34(4), 641-651.
[51] Jiang, J-Q., Cooper, C., Ouki, S., *Chemosphere*, 2002, 47, 711-716.
[52] del Riego, A., Herrero, I., Pesquera, C., Blanco, C., Benito, I., González, F., *Applied Clay Science*, 1994, 9, 189-197.
[53] Zielke, R.C., Pinnavaia, T.J., *Clays and Clay Minerals*, 1988, 36(5), 403-408.
[54] Danis, T., Albanis, T.A., Petrakis, D.E., Pomonis, P.J., *Water Research*, 1998, 32(2), 295-302.
[55] Srinivasan, K.R., Fogler, H.S., *Clays and Clay Minerals*, 1990, 38(3), 277-286.
[56] Srinivasan, K.R., Fogler, H.S., *Clays and Clay Minerals*, 1990, 38(3), 287-293.
[57] Jiang, J-Q., Zheng, Z., *Chemosphere*, 2003, 53, 53-62.
[58] Michot L.J., Pinnavaia, T.J., *Clays and Clay Minerals*, 1991, 39 (6), 634-641.
[59] Bouras, O., Houri, M., Khalaf, H., *Environmental Technology*, 2001, 22, 69-74.
[60] Wu, P.X., Liao, Z.W., Zhang, H.F., Gou, J.G., *Environment International*, 2001, 26, 401-407.
[61] Schoonheydt, R.A., Van Den Eynde, J., Tubbax, H., Leeman, H., Stuyckens, M., Lenotte, I., Stone, W.E.E., *Clays and Clay Minerals*, 1993, 41(5), 598-607.
[62] Ge, Z., Li, D., Pinnavaia, T.J., *Microporous Materials*, 1994, 3, 165-175.
[63] Sarkar, M., Acharya, P.K., Bhattacharya, B., *Journal of Colloid and Interface Science*, 2003, 266, 28-32.
[64] Crank, J., *The Mathematics of Diffusion,* Clarendon, Oxford, 1956.
[65] Bant, F.A, Al-Bashir, B., Al-Asheh, S., Hayajneh, O., Environmental Pollution, 2000, 107, 391-398.
[66] Poots, V., McKay, G., Healy, J.J., *Water Research*, 1976, 10, 1061-1066.
[67] Colleja, G., Serna, J., Rodriguez, J., *Carbon,* 1993, 31(1), 691-697.
[68] Ho, Y.s., McKay, G., *Process Biochemistry*, 1999, 34, 451-465.
[69] Lagergren, S., Zur Theori der Sogenannten Adsorption Gelöster Stoffe, Kungliga Svenska Vetenskapsakademiens, Hondlingar, 1898, 24(4), 1-39.
[70] Sparks, D.L., Kinetics of Soil Chemical Processes, New York, Akademic press, 1989.
[71] Zhang, P-C., Sparks, D.L., *Soil Sci. Soc. Am. J.,* 1993, 57, 3340-345.
[72] Jossens, L., Prausnitz, J.M., Fritz, W., Schlünder, E.U., Myers, A.L., *Chemical Engineering Science,* 1978, 33, 1097-1106.
[73] Halsey, G., Taylor, H.S., *J. Chem. Phys*., 1947, 15, 624.
[74] Song, D-I, Shin, W.S., *Environ. Sci. Technol.*, 39, 1138-1143, 2005.
[75] Lin, S.H., Huang, C.U., Cheng, *J., Environmental Technology*, 2002, 23, 609-622.
[76] Treybal, R.E., *Mass Transfer Operations*, 2nd Edition, McGraw-Hill, New York, 1968.

In: Hazardous Materials in the Soil and Atmosphere
Editor: Robert C. Hudson, pp. 143-160
ISBN 1-60021-250-6
© 2006 Nova Science Publishers, Inc.

Chapter 6

THERMOLYSIS OF TRANSITION METAL COMPLEXES: A REVIEW[1]

Gurdip Singh,* Durgesh Kumar Pandey and Chandra Prakash Singh[2]
Chemistry Department, DDU Gorakhpur University, Gorakhpur, India

ABSTRACT

Thermal behaviour of various transition metal complexes have been reviewed thoroughly. Thermolyses of complexes are studied by TG, DTG, DTA and DSC etc. and pathways of decomposition have been proposed. Any phase transition during thermolysis, studied by the techniques known as 'high temperature reflectance spectroscopy' (HTRS) and 'dynamic reflectance spectroscopy' (DRS) have also been discussed. Intermediates formed during thermolysis has been investigated. Most of the complexes were found to decompose in multisteps. Effect of change of atmosphere on thermal decomposition has also been visualized. End products of thermolysis has also been discussed under oxidative, non-oxidative, quasi-isothermal and isobaric conditions. Stress has been given to compare the thermilytic pattern and thermal stability of the complexes with changing ligand or metal or counter anions. Attention has been given to investigate the kinetics and general mechanism of thermolysis of the complexes.

INTRODUCTION

The union of a metal cation with one or more neutral molecules or ions (known as ligand) is referred as complex compound. The ligand may be monatomic or polyatomic and are capable of independent existence in the absence of coordinate bonding to the central atom. Neutral ligand molecules are H_2O, NH_3, ethylenediamine (en), propylenediamine (pn), pyridine (py), CO, NO, etc. and ions are, anions such as Cl^-, Br^-, NO_3^-, SO_4^{2-}, ClO_4^- or

---

[1] ΨPart: 51

[2] Corresponding author, E-mail: gsingh4us@yahoo.com Phone: 91-551-2200745 (R) 2202856 (O), Fax: 91-551-2340459

cations such as NO^+, $NH_2NH_3^+$, etc. The literature of coordination compounds is so extensive [1] and the number of compounds in this class is very large and diverse having a wide spectrum of applications. Transition metal complexes decompose to respective TMOs and these ultrafine oxides may have interesting electrical, magnetic and catalytic properties [2]. A large number of semiconducting oxides (e.g. CuO, SnO_2, MnO, ZnO, etc) are being explored for application in preparation of environmental sensors for detecting trace level pollutants such as H_2S, CO, NO_x, NH_3, etc. in the atmospheric air [3-8]. When exposed to air, oxygen chemisorbs on the surface of the TMOs to form the species such as O_2^-, O^-, and O^{2-} and influences their surface electrical conductivity. The pollutant gases when present in air react with these species and modulate this conductivity. The change in conductivity is correlated with the pollutant concentration. Hence, by increasing the specific surface area of the TMOs, higher sensitivities can be achieved [9]. Naturally, the nanocrystalline TMOs formed as final decomposition product of the complexes may be better substitute for this application. A good number of pure metal oxides or mixed metal oxides find interesting applications as catalysts in several hydrogenations, dehydrogenation, oxidation and alkylation reactions [10].

The metal complexes consisting of a reducing group (e.g. NH_3, en, pn, etc) as ligand and an oxidizing group (e.g. NO_3^-, ClO_4^-, N_3^-, etc.) as counter anion, exhibit characteristics of energetic compounds. These complexes when raised to sufficient high temperature, whether by direct heating, friction, impact, shock, spark or flame, undergo highly exothermic self-propagative decomposition reaction. The explosive decomposition of such complexes seems to be due to simultaneous presence of both an oxidizing and reducing groups in the same complex molecules. Depending upon the nature of explosivity, these complexes are categorized to lie between those of primary and secondary explosives [11, 12]. On thermolysis, these complexes release the chemical energy available in them and produce gaseous decomposition product and solid residues i.e. respective TMOs [13]. Such complexes may find applications in propellants, explosives and pyrotechnic compositions. Although, TMOs are well known catalysts for burning of CSPs [14-17], many reports are available in open literature which have established that some energetic complexes [18-20] and salts of NTO [21, 22] are better ballistic additives for CSPs and its components than the TMOs. Both heat release on exothermic decomposition of these compounds and final product of decomposition i.e. TMOs were found to be responsible for enhancement of burning rate. All of these additives were studied to be active during condensed phase thermolysis of CSPs and AP. Thus, thermolysis studies on transition metal complexes provide systematic guidance for design and development of various devices and chemical formulations based on such complexes. We are reporting here the thermal behaviour, decomposition pattern and comparative thermal stabilities of complexes of some transition metals.

1. Thermolysis of Cu Complexes

Thermal studies on numerous copper complexes in regards of various facets have been described by good numbers of investigators, which are scattered in literature. Data on various such complexes, obtained after thermal analyses are compiled in Table 1. The effect of various monodentate and bidentate ligands as well as different types of counter anions on thermal stability and decomposition pattern of the complexes are reported. Kinetic analyses have been made based on the data of thermal studies using TG, DTG, DTA/DSC, etc. techniques and pathways of decomposition have been proposed. The general mechanism for thermolysis of Cu complexes may be represented as shown in scheme 1.

$CuL_2X_2.nH_2O$ $\xrightarrow{\text{dehydration}}$ CuL_2X_2 $\xrightarrow{\text{deamination}}$ CuX_2 $\xrightarrow{\text{deanionation and oxidation}}$ CuO

Scheme 1.

Where, x, y and z represent the appropriate numbers of respective species expelled in various decomposition steps, L = ligands, X = counter anion.

Copper (II) usually has coordination number four with a square coplanar geometry [23] and is known for forming well-defined crystalline complexes of different colours with mono or bidentate ligands. The complexes of Cu (II) with ethylenediamine (en) and tetramethylethylenediamine (tmn) show a number of discontinuities during electrical conductivity measurement in solid state at higher temperatures, indicating the structural phase changes on their heating. In order to understand the changes involved, Prabhumirashi and coworkers [24, 25] undertook TG and DTA on a few Cu and Ni complexes. Recently, they [24, 26] have described the thermal properties of various types of Cu (II) complexes of en and tmn. Their study revealed that the thermolysis of such complexes broadly involves three stages i.e. dehydration (in the case of hydrated ones), deamination and deanionation. The dehydration was a slow process and comprises of broad temperature range while deamination was found to be quite steep ranging over an interval of only a few °C. The last step consisting of deanionation with simultaneous oxidation of Cu to CuO was also found to be very broad in nature. However, it was found that the three above stated decomposition processes were not isolated but overlap to varying extent. The overlapping of the process thus causes fractional losses of constituent moieties. These complexes undergo phase changes at lower temperatures, which was associated with the absorption of the required latent heat. The phase change was confirmed by endothermic DTA peak, which was not followed by any actual mass loss in accompanying TG. The dehydration was indicated by broad endothermic peak, deamination by sharp exothermic peak and deanionation by broad exothermic peak in DTA respectively. The broad exotherm for the last process was due to simultaneous occurrence of both deanionation and oxidation of Cu to CuO. However, in the case of [Cu (en) $_2$] $(NO_3)_2$ a steep mass loss followed by a quite steep exothermic DTA peak was observed that was attributed to a chain type highly explosive reaction because of simultaneous presence of an oxidizing group (NO_3) and a reducing group (en) in the same complex molecule. These observations are quite comparable to the studies of Wendlandt [27, 28] *et al*, Nagase [29] *et al*, Srivastava and co-workers [30], and Langfelderova [31] *et al*, made on similar complexes. However, in one of the detailed thermal studies (TG and DTA), reported by Wendlandt [32] on the bis–en complex of Cu (II) SO_4 the thermograms appear to differ somewhat from the thermograms as obtained by Prabhumirashi *et al [16]*. This difference may be attributed to the differences in the atmosphere and heating rates employed. Mathew *et al [12]* have reported the detailed thermal studies on bis(ethylenediamine)copper(II)chloride/bromide monohydrates and some other similar complexes containing both monodentate and bidentate ligands in N_2 atmosphere [33, 34]. Unlike the observations of Prabhumirashi *et al [16]*, they [23] detected the separate mass loss steps for dehydration and deamination.

Table 1. Data from thermal studies and kinetic parameters for various Cu– complexes

Complex	TG Step	Decomposition mode	Ea (kJ/mol) CR	Ea (kJ/mol) HM	DTA peak*	Major temperature (°C)/ nature of peak	ΔH (kJ/mol)	Ea (kJ/mol)
$Cu(en)_2(NO_3)_2$	1	$2en+N_2O_5$	1438	1451	1	65, endo	173	10
					2	240 and 370, exo	328	206
$Cu(en)_2Cl_2{\cdot}H_2O$	1	H_2O	197	211	1	230, exo	149	143
	2	en	133	158	2	470,exo	158	20
	3	en	164	205				
	4	Cl_2	95	128				
$Cu(en)_2SO_4$	1	$2en+0.7SO_4$	143	182	1	190, exo	90	37
	2	$0.3SO_4$	169	188	2	375, exo	795	83
$Cu(en)_2$ $(CH_3COO)_2$ $\cdot H_2O$	1	$H_2O+0.4en$	117	127	1	125, endo	249	22
	2	$1.6en+0.5CH_3COO$	117	146	2	210 and 400, exo	1550	50
	3	$1.5\ CH_3COO$	36	59				
$Cu(en)_2$ $(COO)_2\cdot$ $1.5H_2O$	1	$1.5H_2O+2en+CO_2$	409	234	1	240, exo	925	84
	2	CO	28					
$Cu(tmn)_2$ $(NO_3)_2{\cdot}3H_2O$	1	$2.25H_2O$	191	202	1	75, endo	656	30
	2	$0.75H_2O+2tmn+N_2O_5$	114	140	2	205, exo	1588	351
$[Cu(tmn)Cl_2]_2$	1	1.15tmn	54	63	1	65, endo	115	30
	2	$0.85tmn+0.5Cl_2$	79	98	2	160, exo	693	64
	3	$1.5Cl_2$	386	406	3	425, exo	822	44
$Cu(tmn)_2$ $SO_4{\cdot}4H_2O$	1	$4H_2O + 0.5tmn$	41	62	1	115, endo	235	23
	2	$1.5tmn + 0.7SO_4$	232	268	2	290, exo	565	13
	3	$0.3\ SO_3$	159	179				
$Cu(tmn)$ $(CH_3COO)_2$	1	0.75tmn	94	118	1	185, endo	16	706
	2	$0.25tmn + 2CH_3COO$	842	906	2	220, exo	529	139
$Cu(tmn)$ $(COO)_2\cdot$ $0.5H_2O$	1	$0.5H_2O$	107	121	1	100, endo	72	26
	2	$(tmn + 2CO_2)$	763	778	2	220 and 420, exo	893	83
$[Cu(en)_2]$ $Br_2{\cdot}H_2O$	1	H_2O	195	200	1	91, endo	40	119
	2	en	173	189	2	222, endo	48	117
	3	en + Br						
	4	Volatilization of $CuBr_2$						
$Cu(C_9H_7O_2)_2$	1	C_6H_5			1	223, endo	141	
	2	$C_6H_5+2C_2H_2+CO$			2	286, endo	202	
Cu $(C_6H_{16}N_2)$ $2Cl_2$	1	$Cu_2(C_6H_{16}N_2)3Cl_4$			1	85, endo		
	2	$Cu_2(C_6H_{16}N_2)Cl_4$			2	139, endo		
	3	$CuCl_2$			3	310, endo		
	4	CuO			4	453, exo		
$[Cu(NH_3)_6]$ $(ClO_4)_2$	1	$2NH_3$			1	122, endo		
	2	$4NH_3$+gaseous product			2	222, exo		

*Major DTA peaks for which ΔH and Ea were calculated, exo = exothermic peak, endo = endothermic peak (in DTA

However, in this case also last process i.e. deanionation and volatilization was taking place simultaneously. After dehydration in the first step, one molecule of the en ligand is lost and thus monoligand intermediate is formed [34]. Two endothermic DTA peaks were observed for removal of this first en-molecule, which was due to time lag involved in the breaking of the two coordinate bonds between the bidentate ligand and the metal. It was inferred that in chloride and bromide monohydrate the water molecule occupied in *tetragonal* position, the second being occupied by a halide ion [35, 36]. Hence the thermal decomposition is consistent with the crystal structure of the complex. An intermediate, CuCl was formed at 450 °C and final product was pure Cu at 700 °C, which was confirmed by X-ray powder diffraction [37]. In both the chloride/bromide complexes, the intermediate CuX (X = Cl or Br) is decomposed in final step in which one molecule of en and one chloride/bromide is expelled out [38, 39].

Wendlandt and co-workers [40-42] have published more complete studies of the thermal properties of a number of ethylenediaminetetraacetic acid (EDTA) complexes. Bhat and Krishna Iyer [43] described the thermal decomposition of EDTA complexes of various alkali and transition metal complexes, including that of Copper.

Crespi [44] *et al.* extended the studies of Bhat and Krishna Iyer [43] and described the thermal behaviour and decomposition of some transition metal complexes of EDTA in atmosphere of static air, N_2 and CO_2, which was carried out using TG-DTG and DTA techniques. In the case of copper complex, $H_2[Cu(EDTA)].H_2O$, three steps of decomposition were observed in all the atmosphere used.

Initially the dehydration of the copper complex was observed [44] that was a slow decomposition process as observed earlier by Bhat and Krishna Iyer [43]. Second step comprises of loss of 2(HOOCCH$_2$—N—CH$_2$—CH$_2$) and final step corresponds to loss of other two-carboxyl groups, irrespective of atmosphere used. Although, there is no mention of final residue of thermolysis in N_2 atmosphere, formation of metal and metal oxide (CuO) have been established as residue of thermolysis in atmosphere of CO_2 and air respectively. There is a controversy regarding the final step of mass loss between observations of Crespi *et al* [44] and Bhat *et al* [43], because they [44] reported the loss of two carboxyl groups in final step but such observation was not made by Bhat et al [43].

Paulik and co-workers [45-51] described the combination of titration analysis with derivatograph and the technique was known as *'thermo-gas-titrimetric'* (TGT) in which the gas evolved during thermolysis of a compound is continuously titrated with the help of titrator. Thus, a curve known as TGT curve is obtained. With the help of this technique a study on the gaseous products evolved during thermolysis of $[Cu(NH_3)_4]\ SO_4.H_20$ was made and was found that between 90 and 200 °C one mole of water and two moles of ammonia are evolved simultaneously, and such a complex decomposition process would not be detected by TG and DTG alone. Ammonia is then released stepwise, one mole at a time, but during the release of the fourth mole, sulfur trioxide is also evolved due to the decomposition of copper (II) sulfate. Other applications of this technique include the thermal decomposition of $MnNH_4PO_4{\cdot}H_2O$ [48], $[Ni(NH_3)_6]Cl_2$ [49], $ZnNH_4PO_4$ [49], $CuSO_4{\cdot}5H_20$ [50], and basic iron (III) acetate [50], glucose [50], fructose [50], lactose [50], and $MnCO_3.nH_2O$ [51].

Determination of Kinetic Parameters and Probable Mechanisms for Thermal Decomposition of Cu-Complexes

Kinetic parameters are of great importance for any complex compound, especially for energetic ones, from the application point of view. The enthalpy change (ΔH) and activation energy (E_a) related to various decomposition steps of copper complexes, have been deduced by taking a quantitative study of the thermograms obtained by TG, DTA/DSC etc. Both isothermal and non-isothermal methods have been used to evaluate the kinetics and mechanism of thermal decomposition reactions of various complexes [52-55]. The values of E_a obtained from TG of various copper complexes using Coats and Redfern (CR) [56] and Horowitz and Metzager (HM) [57] equations, are collected in Table 1. Prabhumirashi *et al* [16] have deduced E_a and ΔH involved in thermal decomposition of copper (II) complexes of *en* and *tmn* with various counter anion like NO_3^-, Cl^-, SO_4^{--}, CH_3COO^-, COO^- using DTA technique [58]. The ΔH for a particular process was given by the relation [59, 60] –

$$\Delta H = (k/m)\,A \tag{1}$$

where, A (= $\int \Delta T\, dT$) is area of the corresponding peak in DTA curve, k represents calibration constant and m denotes the mass of the sample.

E_a was given by the equation [60]

$$d\alpha/dt = A\,(1-\alpha)^n\, e^{-E/RT} \tag{2}$$

ΔH and E_a obtained from DTA are listed in Table 1. In the DTA of the complexes, following processes were found to take place

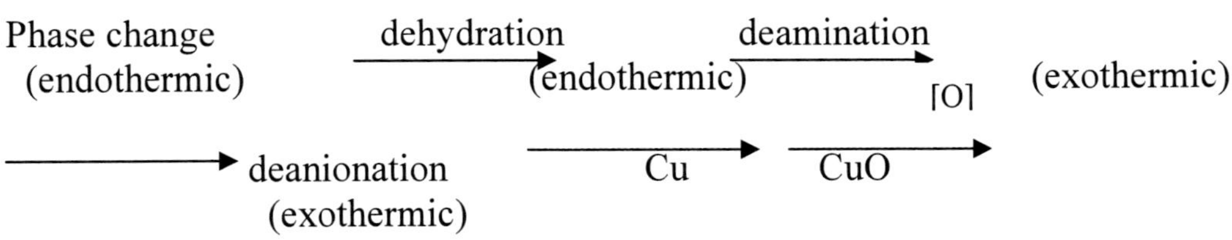

Scheme 2.

The present E_a value for the dehydration of the *en* complex of Cu (II) chloride is of comparable order of magnitude as reported by Langfelderova *et al* [31]. Both ΔH (DTA) and E_a (TG) for dehydration of the *en* complex of Cu (II) oxalate are larger than those for *en* complex of Cu (II) acetate, which might be due to a larger fraction of the ligand lost along with H_2O in the former. The E_a for *en* complex of Cu (II) oxalate was found to be higher than those of *en* complex of any other salt. This may be due to the contribution from the simultaneous loss of ligand and anion part of the ligand.

Higher value of ΔH for removal of *en* as well as *tmn* was observed in the case of Cu (II) oxalate than those of acetate indicating somewhat stronger coordination of these two ligands (*en* and *tmn*) to Cu (II) with oxalate as counter anion than that in the case of acetate. This may be attributable to the chelating bidentate nature of coordination of oxalate against unidentate coordination of acetate. This results in the higher thermal stability of the complex in the case of oxalate. Srivastava *et al.* [30] reported the thermal analysis studies on *en* complex of Cu (II) oxalate, but any thermodynamic or kinetic parameters were not reported.

The ΔH values for the deamination of the *en* complexes of Cu (II) salts were found to increase in the order

$Cl^- < NO_3^- < SO_4^{2-}$

Thus, it implies that the thermal stability increases in the same order. The complexes of Cu (II) chloride are known to be *axially compressed*, while those of Cu (II) nitrate or sulfate are *axially elongated*. It is reported [26] that E_a values for deamination of Cu (II) sulfate are lower than those for complexes of Cu (II) nitrate. This may be due to stronger coordination of SO_4^{2-} with Cu (II) ion than that of NO_3^- and hence weaker coordination of ligands to the Cu (II) ion in presence of SO_4^{2-} as counter ion. The higher value of ΔH in the case of *en* complexes of Cu (II) oxalate and acetate than those for complexes of Cu (II) nitrate, chloride and sulfate could be possible due to the simultaneous loss of ligand and anion in the composite steps in the former.

The probable mechanism for dehydration of majority of the complexes i.e. *en* complexes of Cu (II) chloride, nitrate, oxalate and acetate and the *tmn* complexes of Cu (II) sulfate and oxalate was found [26] to fit more closely to the phase boundary reaction, contracting area cylindrical symmetry mechanism (R_2) [62, 63]. The deamination step in the case of *en* complexes of Cu (II) chloride, nitrate, oxalate and acetate and *tmn* complex of Cu (II) nitrate correspond the mechanism of phase boundary reaction, contracting area cylindrical symmetry (R_2); and that in the case of *tmn* complexes of Cu (II) chloride and sulfate, to random nucleation (F_1). The deanionation in majority of the cases (e.g. *en* complexes of Cu (II) chloride, nitrate, sulfate and oxalate and *tmn* complexes of Cu (II) chloride and acetate) involve the contracting area cylindrical symmetry (R_2) mechanism. However, Mathew *et al.* [23] have inferred that the rate controlling process for both dehydration and deamination of *en* complexes of Cu (II) chloride and bromide monohydrate are random nucleation with the formation of one nucleus on each particle. Similar decomposition mechanism was reported for the decomposition of copper (II) cinnamate complexes [64].

2. THERMOLYSIS OF CO COMPLEXES

Thermolysis of various Co complexes has been studied by a good number of investigators, which have been carried in different atmosphere using TG-DTG and DTA techniques [64-68]. Kinetic parameter of decomposition has been evaluated using mechanistic and non – mechanistic equations. Mechanism of thermolysis has been proposed by identifying intermediates and residues either by using spectral and XRD techniques or from the TG and DTA thermograms. Complexes were found to decompose via different routes in different atmospheres [44]. Wendlandt and coworkers [69] have described the thermal decomposition of $[Co(NH_3)_5H_2O]X_3$ to evaluate the heat of dehydration (ΔH) using sealed tube technique. Simmons and Wendlandt [70] found that the procedural decomposition temperatures for the polymerization reactions of

$[Co(NH_3)_4(H_2O)(CN)Co(NH_3)_4CN](ClO_4)_4$ varied not only with heating rate but also with water content in the surrounding atmosphere.

The *octahedral* → *tetrahedral* structure transition of bis(pyridine) cobalt (II) chloride, $Co(py)_2Cl_2$ has been subject of a number of investigations [71-73]. Wendlandt and George [73] studied the transition using the techniques known as '*high – temperature reflectance spectroscopy'* (HTRS) and *'dynamic reflectance spectroscopy'* (DRS). The thermal transition was found to begin at about 100 and was completed at about 135 °C. The colour change reported for the transition was from violet *(octahedral)* to a dark blue *(tetrahedral)*. It was stated that the change was non – reversible on cooling to room temperature. When 1:1 mass ratio of $[Cr(en)_3]Cl_3$ and NH_4X (X = F, Cl, Br, I, SCN) were heated [74] upto 200 °C in a high - temperature sample holder, the HTRS mode was used to identify the reaction products, while DRS mode was employed to determine the temperature range at which the reaction took place. As a result of these studies, new synthetic procedures were developed for the preparation of *cis* – $[Cr(en)_2X_2]X$, *cis* – $[Cr(pn)_2X_2]X$ and *cis* – $[Cr(tm)_2X_2]X$ complexes (pn = 1,2 – propanediamine, tm = 1,3 – propanediamine). Chang and Wendlandt [75-77] investigated several series of complexes of the types (a) *cis* – and *trans* – $[Co(en)_2(H_2O)_2](NO_3)_2 + NH_4X$ and (b) *cis* – and *trans* – $[Co(NH_3)_4(H_2O)_2](NO_3)_3 + NH_4X$. Thermolysis studies of $[Co(NH_3)_6]Cl_3$ have also been reported by TG – TMA (*thermo-magnetic analysis*) by Simmons and Wendlandt [78]. Patil *et al* [65] have reported the thermal properties of $[Co(NH_3)_6](ClO_4)_2$ and this complex was found to decompose in two steps. In the first step, $4NH_3$ is lost and in second step diammine intermediate decomposes explosively to CoO and not to $CoCl_2$. These observations are analogous to those reported earlier [79-81]. Maslowska and Szmich [67] have carried out TG and DTA on Co (II) complexes of thiosalicylic acid (SR) and ethylenediamine (en) like, [Co(SR) and $[Co(en)_2](NO_3)_2$ in static air atmosphere at a heating rate of 5 °C min^{-1}. They found that both the complexes decompose in multistep and cause endo effect in DTA. Finally, CoO was obtained as decomposition product. Wendlandt [82] has shown that thermal stabilities of salts containing the $[Co(NH_3)_6]^{3+}$ ion (heated in helium at 5 °C min^{-1}) vary with the anion present (Cl^-, Br^-, NO_3^- and SO_4^{2-}) and that these anions also influence the chemical changes which occur.

The sulfate, $[Co(NH_3)_6]_2(SO_4)_3 \cdot 2.5H_2O$, was shown by TG to be stable to about 220 °C, but a sequence of at least seven endotherms, accompanied by gas evolution, was detected. $[Co(NH_3)_6](NO_3)_3$ was reported to decompose exothermically in single step and gaseous products of decomposition were in order of abundance: H_2O, N_2, NH_3 and N_2O. Watt [83], using a slower heating rate, identified the process at 280 °C as conversion of the $[Co(NH_3)_6]Cl_3$ to $[Co(NH_3)_5Cl]Cl_2$. The initial step in decomposition was, therefore, identified as electron transfer [84] giving the less stable Co (II) coordination compound. $(NH_4)_2CoCl_4$ was formed as intermediate, which represents a novel alternative to other proposed mechanism [83, 85]. On further heating, the intermediate dissociates to leave a residue. In vacuum, deammination is enhanced, the reaction temperature is slightly decreased and substitution proceeds a step further to *trans* - $[Co(NH_3)_4Cl_2]Cl$ [86].

Zivkovic [87] using heating rate between 2.5 and 20 °C min^{-1} found E_a values for decomposition of $[Co(NH_3)_6]Cl_3$ to $CoCl_2$ and $(NH_4)_2CoCl_4$ (180 to 280 °C) and the further decomposition of $(NH_4)_2CoCl_4$ to NH_4Cl and $CoCl_2$ (350 °C) to be 132 and 54 to 67 kJ mol^{-1}, respectively.

The behaviour of bromide, $[Co(NH_3)_6]Br_3$ is analogous to that given for chlorides, though the reaction temperature may be slightly different and in vacuum, $[Co(NH_3)_5Br]Br_2$ probably [85] does not yield *trans* - $[Co(NH_3)_4Br_2]Br$.

Wendlandt and Smith [88] also studied the reaction of pentamminehalocobalt (III) nitrates, $[Co(NH_3)_5X](NO_3)_2$ using TG between 127 and 177 °C. They found that the stability sequence for the initial dissociation step was (X =) $F^- = Cl^- > Br^- > I^-$.

Joyner [89] has described the isothermal decomposition of solid hexaammine–, azidopentaammine, and *cis*– and *trans*– diazidotetraammine cobalt (III) azide. This study was consistent with the influence of ammonia and residual products on reaction rates and was supplemented by optical microscopy and X-ray identification of the phases present. Hexammine cobalt (III) azide $[Co(NH_3)_6](N_3)_3$ was found to decompose [89].

$$[Co(NH_3)_6](N_3)_3 \rightarrow CoN\ (s) + 6NH_3\ (g) + 4N_2\ (g)$$

$$[Co(NH_3)_6](N_3)_3 \rightarrow Co(NH_3)_2(N_3)_2\ (s) + 4NH_3\ (g) + 1.5N_2\ (g)$$

Scheme 3.

These processes will be referred to as the CoN and cobalt (II) reactions. Analogous reactions may be written for the penta– and tetraammine.

By making use of TG, DTA and evolved gas analysis (EGA), Wendlandt and Smith [90] identified the two intermediates in the dissociation of $[Co(NH_3)_6]Cl_2$

$$[Co(NH_3)_6]Cl_2 \longrightarrow [Co(NH_3)_2]Cl_2 + 4NH_3$$
(TG 87 to 187 °C, DTA endo. peak at 143 °C K)

$$\longrightarrow [Co(NH_3)Cl_2] + NH_3$$
(TG 187 to 220 °C, DTA endo. peak at 231 °C)

$$\longrightarrow CoCl_2 + NH_3$$
(TG 220 to 300 °C, DTA endo. peak 286 °C)

Scheme 4.

The behaviour of bromide was similar to that of chloride. In the case of iodide, $[Co(NH_3)_2I_2]$ was obtained as intermediate at about 177 °C. This compound melted at about 207 °C and decomposes at about 297 °C to form CoI_2 (I- was not oxidized).

Collins *et al* [91] concluded (after using TG, DTA, magnetic, spectroscopic measurement and mass spectra) that, in nitrogen, the chloride $[Co(en)_3]Cl_3$ and the analogous bromide decompose as follows:

$$[Co(en)_3]Cl_3 \longrightarrow \textit{trans}\text{-}[Co(en)_2Cl_2]Cl \longrightarrow CoCl_2 + (NH_4)_2CoCl_4 \longrightarrow CoCl_2 + 2NH_4Cl$$

Scheme 5.

However, Zsako and coworkers [92] concluded that the formation of $[Co(en)_2X_2]X$ is only possible when X = Cl^-, because other ligands studied (Br^-, I^-, NO_3^-, CNO^-) are susceptible to reactions involving the Co^{3+} ion.

Table 2. Some data from thermal studies of various Co – complexes

Complex	TG				DTA		
	Step	Expelled moiety	mass loss obsd.	Calc.	Peak temp. (°C)	Nature of peak	Reference
$[Co(NH_3)_6](ClO_4)_2$	1	$4NH_3$	18.78	18.89	122	endo	102
	2	Explosion products			222	exo	
$Co(C_9H_7O_2)2{\cdot}2H_2O$	1	$2H_2O$	9.84				101
	2	$2C_6H_5$+2CO	54.19				
	3	$2C_2H_2$ + $1/2O_2$	18.16				
	4	addition of $1/2O_2$	+4.95				
	5		+2.95				
$Co(C_6H_{16}N_2)_2Cl_2$	1	* $Co_3(C_6H_{16}N_2)_5Cl_6$	10.6	10.4	76	endo	103
	2	$Co(C_6H_{16}N_2)Cl_2$	21.3	20.8	132	endo	
	3	$CoCl_3$	31.6	31.2	248	endo	
	4	Co_3O_4	15.0	16.0	520	exo	
Co(SR) SR = thiosalicylate	1		19			endo	104
	2		26			endo	
	3	Oxidation products	14			exo	
	4		4			endo	
$[Co(en)_2](NO_3)_2$	1	2en	33			Endo	104
	2	Decomposition of $Co(NO_3)_2$	41			endo	
	3		3			endo	
$H_2[Co(EDTA)]{\cdot}2H_2O$							
N_2 atmosphere	1	$2H_2O$	9.55	9.36	140	endo	81
	2	$4CH_2 + H_2O$	19.10	19.25	250	endo	
	3	2(NCOO)	30.01	30.12	390	endo	
	4		25.23	25.98	680	exo	

Table 2. Some data from thermal studies of various Co – complexes (Continued)

Complex	TG Step	Expelled moiety	mass loss obsd.	Calc.	DTA Peak temp. (°C)	Nature of peak	Reference
CO_2	1	$2H_2O$	9.78	9.36	200	endo	81
	2	$CH_2 + H_2O$	8.38	8.32	310	endo	
	3	$3CH_2$	10.47	10.93	450	endo	
	4	3(NCOO)	30.03	3012	520	endo	
	5		25.84	25.98			
Air	1	$2H_2O$	9.82	9.36	150	endo	81
	2	$4CH_2 + H_2O$	18.93	19.25	450	exo	
	3	pyrolytic product of intermediate	49.09	49.87	920	endo	
$[Co(NH_3)_6]_2(C_2O_4)_3{\cdot}4H_2O$	1	*$[Co(NH_3)_6]_2(C_2O_4)_3$	10.8	10.9	183	endo	
	2	$CoC_2O_4{\cdot}NH_3$	40.5	39.2	231	endo	
	3	CoC_2O_4	2.5	5.2	287	endo	
	4	$CoO + CoC_2O_4$ (75 mol%)	6.2	5.5	338	endo	
	5	Co + CoO (50 mol %)	20	18.8	385	endo	

*resulting intermediate after decomposition of parent compound, endo = endothermic peak, exo = exothermic peak (in DTA)

*Major DTA peaks for which ΔH and Ea were calculated, exo = exothermic peak, endo = endothermic peak (in DTA

Fakiha *et al* [64] conducted TG – DTG and DSC on cobalt (II) cinnamate complex like $Co(C_9H_7O_2)_2{\cdot}2H_2O$ in N_2 atmosphere and inferred that complete decomposition of the complex involves five stages as reported in Table 2. Spectroscopic and magnetic data [93] predict that this complex has a planar arrangement with the carboxylate group of two different cinnamate ions bonded to a metal atom to give one layer that makes the C_6H_5 ring free out of the plane of the layer, thus facilitating its decomposition. Prolong heating of the Co (II) complex up to 730 °C causes interaction of the residue with oxygen, forming Co_2O_3 oxide. Kinetic parameters were evaluated using C – R method and three-dimensional diffusion D_3 was found to operate as decomposition mechanism.

Allan *el at* [66] have described the thermal decomposition mode of chloro complex of cobalt with 1,6 – hexanediamine such as $Co(C_6H_{16}N_2)_2Cl_2$ using TG and DTA at a heating rate of 6 °C min^{-1} in air. The pink coloured cobalt complex undergoes endothermic reaction to give first a blue complex, $Co_3(C_6H_{16}N_2)_5Cl_6$ and then a green complex, $Co(C_6H_{16}N_2)Cl_2$. This is followed by a further endothermic reaction to give cobalt chloride and then an exothermic reaction to give Co_3O_4.

Ingier–Stocka and Maciejewski [68] have studied the thermolysis of $[Co(NH_3)_6]_2(C_2O_4)_3{\cdot}4H_2O$ under non – isothermal, isothermal and quasi – isothermal – isobaric conditions in air and argon atmosphere. They concluded that first step of decomposition is dehydration, followed by decomposition of anhydrous salt and reduction of Co (III) to Co (II) in the second stage. The third stage in air is decomposition of

$Co[Co(C_2O_4)_2]$ to Co_3O_4 and in argon, $Co[Co(C_2O_4)_2]$ decomposes to metallic cobalt and cobalt (II) chloride. The end product under quasi – isothermal – isobaric condition is metallic cobalt. The agent reducing Co (III) to Co (II) is ammonia [94, 95]. They [68] have also reported [96-99] the mechanism of thermal decomposition of hexaamminecobalt(III)chloride and hexaamminecobalt(III)oxalate tetrahydrate.

3. THERMOLYSIS OF NI COMPLEXES

Nickel forms well defined crystalline complexes of different colours with various monodentate, chelating or polydentate ligands. Thermolysis of these complexes has been studied by variety of thermoanalytical techniques such as TG, DTG, DTA, DSC, HTRS – DRS etc. Paulik and Paulik [45] found that the thermal decomposition of $Ni(NH_3)_6Cl_2$, dissociated in three separate steps, at 180, 320 and 360 °C, respectively. A slight overshoot of the curve during the first dissociation reaction was observed which was due to an induction period caused by delayed nucleus formation.

Yang [100] studied the deamination of $Ni(py)_4Cl_2$ by using HTRS, and the spectrum at 25 °C was that for initial compound, $Ni(py)_4Cl_2$. From 125 to 175 °C, two moles of pyridine per mole of complex are lost, so that the spectrum at 175 °C was that for the complex $Ni(py)_2Cl_2$. From 175 to 275 °C, another pyridine is evolved, so that the 275 °C spectrum was that for $Ni(py)Cl_2$.

During the thermolysis of $[Ni(NH_3)_6]Cl_2$, by TG and DTA, it was observed [27] that four molecules of NH_3 are lost at about 85 °C and remaining two molecules of NH_3 are removed at about 215 °C. $[Ni(NH_3)_6]SO_4$ gave some evidence for diamine production, which further reacted, to $NiSO_4$ by about 527 °C. $[Ni(NH_3)_6](NO_3)_2$ lost 2 or 3 molecules of ammonia prior to an exothermic oxidation reaction (260 °C) [100]. NH_3 is replaced by NO_3^- during very slow oxidation of Ni^{2+} while maintaining the *octahedral* coordination of the central atom. The complexity of the oxidation reaction involving nitrate decomposition in these coordination is shown by a study which concluded that the analysis of the product of the reaction of $[Ni(aniline)_2](NO_3)_2{\cdot}4H_2O$. In addition to H_2O and NH_3, some 44 different organic products were detected.

Sherriff and Galway [102] detected that the first step of decomposition of hexaamminenickel(II)perchlorate is the reversible dissociation

$$[Ni(NH_3)_6](ClO_4)_2 \leftrightarrow [Ni(NH_3)_2](ClO_4)_2 + 4NH_3$$

Similar dissociation of this nickel complex is also observed by Patil *et al* [65]. Isothermal $\propto$ – time curves showed an induction period followed by a deceleratory reaction, which could be described by the contracting volume equation with a discontinuity at $\propto \approx 0.35$. The value of E_a (71 kJ mol^{-1}) for this step was close to the enthalpy of dissociation and the apparent energy barrier to the diffusion controlled reammination reaction was small ($\approx$ 8 kJ mol^{-1}).

Jona *et al.* [103] carried out a similar study on the dissociation of various coordination compounds like $[NiL_2(NCS)_2]$ (L = pyridine, α– or β– picoline, 2,6 – lutidine or quinoline). They established that the nature of the ligand influenced whether the reactions occurred in one or in two steps, and that reaction also depended on the physical conditions. The same workers [104] have shown that the sequence of steps involved in dissociation of $[Ni(NH_3)_4](NCS)_2$ was

$$[Ni(NH_3)_4](NCS)_2 \longrightarrow [Ni(NH_3)_3](NCS)_2 \longrightarrow [Ni(NH_3)_2](NCS)_2 \longrightarrow Ni(NCS)_2$$

Scheme 6.

The dissociation reactions of ethylenediamine coordination compounds of nickel probable [105] involved the evolution of a single en molecule as the first step in salt breakdown

$$[Ni(en)_3]Br_2 \longrightarrow [Ni(en)_2]Br_2 + en$$

The phase transition in $[Ni(en)_2(NCS)_2]$ single crystal was studied [106] using a white beam synchrotron radiation. The phase transition commenced at defect sites. The role of generation and relaxation of mechanical stresses in influencing kinetics of solid-state transformations was later reviewed [107].

Maslowska *et al* [67] studied decomposition of the [Ni(SR)] and $[Ni(en)_3](NO_3)_2$ in static air using TG and DTA at a heating rate of 5 °C min^{-1}. The complexes were found to decompose in multisteps and NiO was obtained as final residue. en complex was relatively more thermally stable than SR counterpart. Similar results were obtained when TG – DTG and DSC was carried out on $Ni(C_9H_7O_2)_2{\cdot}2H_2O$ in which three dimensional diffusion (D_3) mechanism was found to be involved [64].

The blue nickel complex, $Ni(C_6H_{16}N_2)_2Cl_2$, undergoes [66] an endothermic reaction to give a green complex, $Ni(C_6H_{16}N_2)Cl_2$. A subsequent endothermic reaction then takes place, followed by exothermic reactions to give nickel chloride and NiO, respectively.

$H_2[Ni(EDTA)]{\cdot}2.5H_2O$ was studied to decompose via different routes in N_2, CO_2 and air atmosphere [44]. In N_2, the complex decomposes in four steps, and in two steps in CO_2 and air. Generally, first step corresponds to dehydration; second decomposition of anhydrous complex and finally decomposition takes place to yield NiO. The bromide complex of nickel (II) with 1,3 –dipiperidinopropane i.e. $Ni[CH_2(CH_2NC_5H_{10})_2]Br_2$ showed two steps of mass losses upon heating up to 700 °C [107], with simultaneous loss of organic and inorganic fragments. The residue after heating corresponds to metal oxide.

4. Thermolysis of Zn Complexes

A study of the thermal stability and thermal decomposition of $H_2[Zn(EDTA)]{\cdot}2H_2O$ was made by Crespi et al [44] using TG – DTG and DTA in N_2, CO_2 and air atmospheres. The complex was detected to decompose in four steps in N_2 and in five steps both in CO_2 and air atmospheres. These decomposition steps generally corresponds to dehydration in first step, removal of one or more CH_2 and H_2O molecules in second step, loss of CH_2NCOO and NCO in third and finally decomposition takes place to give ZnO.

Zinc (II) bromide complex with 1,3 – dipiperidinopropane [108] is decomposed in two steps finally giving ZnO as residue. The TG curve of this complex indicated the occurrence of processes causing an increase in the mass of residue of the expected product of the decomposition in air is the oxides [109-113]. TG, DTA and DTG curves for this complex showed that the release of volatile species occur in the two steps process according to the following equation

(i) $[ZnCH_2(CH_2NC_5H_{10})_2Br_2] \longrightarrow [Zn(CH_2NC_5H_{10})Br_2] + CH_2CH_2NC_5H_{10}$

$[Zn(CH_2NC_5H_{10})Br_2] + (1/2)\ O_2$

$ZnO + CH_2NC_5H_{10} + Br_2$

$\longrightarrow$

Scheme 7.

The formation of ZnO has been confirmed by chemical analysis and IR spectroscopy.

Zelenak *et al.* [114] have reported the thermolysis studies on various zinc (II) complexes containing theophylline like $[Zn(NH_3)_2(Tph^-)_2]$ (Tph^- = theophyllinato monoanion), $[Zn(mea)_2(Tph^-)_2]$ (mea = ethanolamine), $[Zn(ipa)_2(Tph^-)_2]$ (ipa = isopropylamine) and $[Zn_2(OH)(CH_3COO)\ (Tph^-)_2]$ using TG, DTG and DTA. The complexes were detected to decompose in multisteps and as expected, ZnO was obtained as product of thermolysis. The thermal stabilities of these complexes increases in the order:

$[Zn(mea)_2(Tph^-)_2] < [Zn(NH_3)_2(Tph^-)_2] < [Zn(ipa)_2(Tph^-)_2]$

180 °C 190 °C 205 °C

$< [Zn_2(OH)(CH_3COO)\ (Tph^-)_2]$

360 °C

Thermal properties of the theophylline complexes of Co (II), Ni (II) and Cu (II) have also been investigated [115, 116].

Maslowska *et al.* have reported thermal decomposition of the [Zn(SR)] (SR = thiosalicylate) and $[Zn(en)_2](NO_3)_2 \cdot 2H_2O$ in static air using TG and DTA at a heating rate of 5 °C min^{-1}. The complexes were found to decompose in multisteps. en complex was relatively less thermally stable than SR counterpart. [Zn(SR)] is decomposed exothermically to ZnO while all the steps of decomposition of $[Zn(en)_2](NO_3)_2 \cdot 2H_2O$ gave endothermic peaks in DTA and ZnO was the final product of thermolysis.

5. Applications of Metal Complexes in Pyrotechnics

The substances used in the past in pyrotechnic composition are now avoided since they lead to environmental and health problems. Some alternative metal complexes have been synthesized to replace the conventional oxidizer like lead azide, potassium nitrate etc in view of their inherent drawbacks such as friction sensitivity, lesser stability to moisture and incompatibility with other materials used in ammunition. Nickel hydrazine nitrate [117] has been prepared from available raw materials and found insensitive to impact, friction or electrostatic charge but more sensitive to flame [118]. It is found to be a suitable replacement for lead azide as an intermediate charge in commercial detonators. Also, the reaction liquor, unlike lead azide, does not contain any useless ions and thus makes possible the repeated use of the liquor in continuous preparation of this primary explosive. N, N-bis(2,4,6 – trinitrobenzoyl)lead hydrazine, synthesized from lead acetate and bis(2,4,6 – trinitrobenzoyl)hydrazine with better heat resistant properties is found to be a good substitute for lead styphnate [119]. Cobalt perchlorate complex with diaminourea was found to be good initiator for use as primary explosive [120].

REFERENCE

[1] G. Wilkinson, R.D. Gillard and J.A. Mc Cleverty (Eds.), *Comprehensive Coordination Chemistry*, Pergamon, Oxford, 1987, Vols 1 – 7.

[2] S.Y. Sawant, K.R. Kannan and V.M.S. Verenkar, in: *Proceedings of Thirteenth National Symposium on Thermal Analysis, Thermans 2002*, pp 154.

[3] P.T. Moseley and D.E. Williams, *Sensors and Actuators B*, 1 (1990) 113.

[4] N. Yamazoe and N. Miura, in: *Chemical Sensors Technology*, Vol. 4 (Ed. S. Yamaguchi) Elsevier, 1992, 19.

[5] D.E. Williams, *Anal. Proc.*, 28 (1991) 366.

[6] G. Mangamma, V. Jayaraman, T. Gnanasekaran and G. Periaswami, *Sensors and Actuators B*, 53 (1999) 133.

[7] V. Jayaraman, K.I. Gnanaseker, E. Prabhu, T. Gnanasekaran and G. Periaswami, *Sensors and Actuators B*, 54 (1999) 147.

[8] V. Jayaraman, K.I. Gnanaseker, E. Prabhu, T. Gnanasekaran and G. Periaswami, *Sensors and Actuators B*, 54 (1999) 175.

[9] R. Sridharan, V. Jayaraman, R. Ganesan and T. Gnanasekaran, in: *Proceedings of Thirteenth National Symposium on Thermal Analysis, Thermans 2002*, pp 15.

[10] R. Prasad, in: *Proceedings of Thirteenth National Symposium on Thermal Analysis, Thermans 2002*, pp 106.

[11] W. Friendrich and P. Vervoost, *CA* 21 (9124) 1184.

[12] K.C. Patil and V.R. Pai Verneker, *Combust. Flame.*, 25 (1978) 387.

[13] G. Singh, I.P.S. Kapoor, S.M. Mannan and J. Kaur, *J. Hazard. Mater.*, A 79 (2000) 1.

[14] P.W.M. Jacobs and H.M. Whitehead, *Chem. Rev.*, 69 (1969) 551.

[15] H.H. Kung, *Transition Metal Oxides, Surface Chemistry and Catalysts*, Vol. 45, Elsevier, Amsterdam, 1998.

[16] Shin – Ming Shen, Sun – I – Chen and Bur – Horng Wu, *Thremochim. Acta*, 223 (1993) 135.

[17] F. Solymosi, *Structure and Stability of Salts of Halogen Oxiacids in the Solid Phase*, John Wiley and Sons, New York, 1977, pp 195.

[18] K.C. Patil and V.R. Pai Verneker, *Combust. Flame*, 25 (1975) 387.

[19] G. Singh and D.K. Pandey, *J. Energ. Mater.*, In Press, 2002.

[20] G. Singh and D.K. Pandey, *J. Indian Chem. Soc.*, 2002, communicated.

[21] G. Singh, I.P.S. Kapoor, S.K. Tiwari, S.P. Felix, K.N. Ninan and T.L. Varghese, *J. Energ. Mater.*, In Press, 2002.
[22] G. Singh and S.P. Felix, *Combust. Flame*, In Press, 2002.
[23] S. Mathew, C.G.R. Nair and K.N. Ninan, *Thermochim. Acta*, 181 (1991) 253.
[24] L.S. Prabhumirashi, G.N. Natu and J.K. Khoje, *J. Therm. Anal.*, 35 (1989) 1097.
[25] L.S. Prabhumirashi, G.N. Natu and S.R. Nayak, *J. Therm. Anal.*, 35 (1989) 1105.
[26] L.S. Prabhumirashi and J.K. Khoje, *Thermochim. Acta*, 383 (2002) 109.
[27] T.D. George and W.W. Wendlandt, *J. Inorg. Nucl. Chem.*, 25 (1963) 395.
[28] W.W. Wendlandt, *The Thermal Properties of Transition Metal Amine Complexes*, Elsevier, Amsterdam, 1967 (Chapter 2).
[29] K. Nagase and H. Yokobayash, *Bull. Chem. Soc. Jpn.*, 47 (1974) 2036.
[30] P.C. Srivastava, B.N. Singh, S.D. Adhya and K.C. Banerji, *J. Therm. Anal.*, 27 (1983) 263.
[31] H. Langfelderova, V. Karla, M. Lirkesova and J. Gazo, *J. Therm. Anal.*, 26 (1983) 95.
[32] W.W. Wendlandt, *J. Inorg. Nucl. Chem.*, 25 (1963) 833.
[33] S. Mathew, C.G.R. Nair and K.N. Ninan, *Thermochim. Acta*, 144 (1989) 33.
[34] C.G.R. Nair, S. Mathew and K.N Ninan, *Thermochim. Acta*, 150 (1989) 63.
[35] F. Mazzi, *Rand. Soc. Mineral. Ital.*, 9 (1953) 148.
[36] R.D. Ball, D. Hall, C.E.F. Pickard and T.N. Waters, *J. Chem. Soc. A*, 1967, 1435.
[37] *Hannewalt Method Search Manual, International Centre for Diffraction Data*, JCPDS, Swarthmore, PA pp 748.
[38] J.E. House, Jr. and F.M. Tahir, *Thermochim. Acta*, 117 (1987) 371.
[39] J.R. Allan, B.R. Carson, D.L. Gerrard and J. Birnie, *Thermochim. Acta*, 160 (1990) 329.
[40] W.W. Wendlandt, *Anal. Chem.*, 32 (1960) 848.
[41] W.W. Wendlandt and G.R. Horton, *Nature*, 187 (1960) 769.
[42] M.L. Morris, R.W. Dunham and W.W. Wendlandt, *J. Inorg. Nucl. Chem.*, 20 (1961) 274.
[43] T.R. Bhat and R. Krishna Iyer, *J. Inorg. Nucl. Chem.*, 26 (1967) 179.
[44] M.S. Crespi, C.A. Ribeiro and M. Ionashiro, *Thermochim. Acta*, 221 (1993) 63.
[45] J. Paulik and F. Paulik, *Thermochim. Acta*, 4 (1972) 189.
[46] J. Paulik, F. Paulik and L. Erdey, *Mikrochim. Acta*, (1966) 886.
[47] J. Paulik and F. Paulik, *Talanta*, 17 (1970) 1224.
[48] J. Paulik and F. Paulik, in: *Proceedings of the Third Analytical Chemistry Conference*, Budapest, 1970, pp 225.
[49] J. Paulik and F. Paulik, in: *Proceedings of the Third Analytical Chemistry Conference*, Budapest, 1970, pp 231.
[50] S. Gal, J. Simon and L. Erdey, in: *Proceedings of the Third Analytical Chemistry Conference*, Budapest, 1970, pp 243.
[51] F. Paulik and J. Paulik, *Thermochim. Acta*, 3 (1971) 17.
[52] C. De, P.K. Biswas and N.R. Chaudhuri, *Bull. Chem. Soc. Jpn.*, 56 (1983) 3145.
[53] J.E. House, Jr., and F.M. Tahir, *Thermochim. Acta* 118 (1987) 191.
[54] J. Ribas, M,serra, A. Escuer and H.D. Baro, *Thermochim. Acta*, 80 (1984) 103.
[55] J.E. House. K.A. Kemper and H.M. Fogel, *Thermochim. Acta*, 129 (1988) 263.
[56] A.W. Coats and J.P. Redfern, *Nature*, 201 (1964) 68.
[57] H.H. Horowitz and G. Metzger, *Anal. Chem.*, 35 (1963) 1464.

[58] W.W. Wendlandt, *Thermal Methods of Analysis, 2nd Edition*, Wiley, New York, 1973, (Chapter 5).
[59] S. Speil, L.H. Berkelhammer, J.A. pask and B. Davis, *Bur. Mines Tech. Papers*, 1945, 664.
[60] P.F. Kerr and J.L. Kulp, *Am. Minerlogist*, 33 (1948) 387: CA 42537i.Blazek, *Thermal Analysis*, Van Nostrand Reinhold, London, 1973 (Chapter 2).
[61] D. Dollimore and C.J. Keattch, *An Introduction to Thermogravimetry*, 2nd Edition, Heyden, London, 1975 (Chapter 5).
[62] W.W. Wendlandt, *Thermal Methods of Analysis*, 2nd Edition, Wiley, New York, 1973, (Chapter 2).
[63] S.A. Fakiha, K.H. Halawani and K.M. Abd El – Salaam, *J. Therm. Anal.*, 39 (1993) 913.
[64] K.C. Patil and V.R. Pai Verneker, *Thermochim. Acta*, 15 (1976) 257.
[65] J.R. Allan, P.C. Beaumont, G.H.W. Milburn and I. Wood, *Thermochim. Acta*, 219 (1993) 159.
[66] J. Maslowska and J. Szmich, *J. Therm. Anal*, 36 (1990) 2701.
[67] Ewa Ingier – Stocka, M Maciejewski, *Thermochim. Acta*, 354 (2000) 45.
[68] W.W. Wendlandt, G. D'Ascenzo and R.H. Gore, *J. Inorg. Nucl. Chem.*, 32 (1970) 3404.
[69] E.L. Simmons and W.W. Wendlandt, *Thermochim. Acta*, 2 (1971) 465.
[70] W.W. Wendlandt and T.D. George, *Chemist – Analyst*, 53 (1964) 100.
[71] W.W. Wendlandt, *J. Therm. Anal.*, 1 (1970) 469.
[72] W.W. Wendlandt and T.D. George, *Chemist – Analyst*, 53 (1964) 71.
[73] W.W. Wendlandt and C.H. Stembridge, *J. Inorg. Nucl. Chem.*, *27* (1965) 575.
[74] F.C. Chang and W.W. Wendlandt, *J. Inorg. Nucl. Chem.*, 32 (1970) 3535.
[75] F.C. Chang and W.W. Wendlandt, *Thermochim. Acta*, 2 (1970) 293.
[76] F.C. Chang and W.W. Wendlandt, *Thermochim. Acta*, 3 (1971) 69.
[77] E.L. Simmons and W.W. Wendlandt, *Anal. Chim. Acta*, 35 (1966) 461.
[78] W. Friendrich and P. Vervoost, *CA*, 21 (1927) 1184.
[79] W.W. Wendlandt and J.P. Smith, *J. Inorg. Nucl. Chem.*, 25 (1963) 985.
[80] T.M. Southern and W.W. Wendlandt, *J. Inorg. Nucl. Chem.*, 32 (1970) 3783.
[81] W.W. Wendlandt, *J. Inorg. Nucl. Chem.*, 25 (1963) 545.
[82] G. W. Watt, *Inorg. Chem.*, 3 (1964) 325.
[83] N. Tanaka and M. Nanjo, *Bull. Chem. Soc. Jpn*, 37 (1964) 1330.
[84] E.L. Simmons and W.W. Wendlandt, *Anal. Chim. Acta*, 35 (1966) 461; *J. Inorg. Nucl. Chem.*, 28 (1966) 2187.
[85] L.W. Collins, W.W. Wendlandt and E.K. Gibson, *Thermochim. Acta*, 8 (1974) 303.
[86] Z.D. Zivkovic, *J. Therm. Anal.*, 41 (1994) 99.
[87] W.W. Wendlandt and J.P. Smith, *J. Inorg. Nucl. Chem.*, 25 (1963) 843,1267; 26 (1964) 445.
[88] T.B. Joyner, *J. Phy. Chem.*, 69 (1965) 1723; 71 (1967) 3431; 72 (1968) 703, 4386; 74 (1970) 1552, 1558, 1563.
[89] W.W. Wendlandt and J.P. Smith, *J. Inorg. Nucl. Chem.*, 25 (1963) 985.
[90] L.W. Collins, W.W. Wendlandt and E.K. Gibson, *Thermochim. Acta,* 8 (1974) 205, 307.
[91] J. Zsako, Cs. Vashelyi, G. Liptay and K. Szilagyi, *J. Therm. Anal.*, 7 (1975) 41.

[92] J.R. Allan, B.R. Carson, D.L. Gerrard and S. Heoy, *Thermochim. Acta*, 154 (1989) 315.
[93] W.W. Wendlandt and J.P. Smith, *The Thermal Properties of Transition Metal Amine Complexes*, Elsevier, Amsterdam, 1967, pp 37 – 103, 43.
[94] B.N. Figgis and J. Lewis, *Magnetochemistry in Technique of Inorganic Chemistry*, Wiley, New York, 1965, pp 142.
[95] E. Ingier – Stocka and A. Bogacz, *J. Therm. Anal.*, 35 (1989) 1373.
[96] E. Ingier – Stocka, *J. Therm. Anal.*, 37 (1991) 521, 769.
[97] E. Ingier – Stocka, *Thermochim. Acta*, 170 (1990) 107.
[98] E. Ingier – Stocka, *J. Therm. Anal.*, 50 (1997) 603.
[99] W.Y. Yang, Unpublished results.
[100] M. Prost, E. Garbowski and M.V. Mathieu, *J. Chim. Phys.*, 71 (1974) 1519.
[101] R.A.F. Sherriff and A.K. Galwey, *J. Chem. Soc. A*, (1967) 1705.
[102] E. Jona, T. Sramko and J. Gazo, *J. Therm. Anal.*, 4 (1972) 61.
[103] E. Jona, T. Sramko and J. Gazo, *J. Therm. Anal.*, 7 (1975) 551.
[104] W.W. Wendlandt and J.P. Smith, *The Thermal Properties of Transition Metal Amine Complexes*, Elsevier, Amsterdam, 1967, pp 37 – 103, 43.
[105] T.P. Shahtshneider, E. Yu. Ivanov and V.V Boldyrev, *React. Solids*, 2 (1087) 339.
[106] A.P. Chupakhin, A.A. Sidelnikov and V.V. Boldyrev, *React. Solids*, 3 (1987) 1.
[107] S. Rehman, M. Arshad, S.A. Khan, K. Masud, N. Arshad, A.H. Qureshi and S.A. Ghauri, *Thermochim. Acta*, 345 (2000) 81.
[108] Bujewski, K. Grzedzicki and Z. Warnke, *J. Therm. Anal.* 33 (1988) 961.
[109] Tossidis and A. Syngollitou – Kourakou, *J. Therm. Anal.*, 32 (1987) 491.
[110] J. Maslowska and A. Baranowska, *J. Therm. Anal.*, 29 (1984) 309.
[111] J. Cernak and J. Chomic, *J. Therm. Anal.*, 32 (1987) 527.
[112] B.N. Sivasankar and S. Govindarajan, *J. Therm. Anal.*, 46 (1996) 117.
[113] V. Zelenak, K. Gyoryova and E. Andogova, *Thermochim. Acta*, 354 (2000) 81.
[114] P. Bombicz, J. Madarasz, E. Forizs and I. Foch, *Polyhedron*, 16 (1997) 3601.
[115] S. Gal, J. Madarasz, E. Forizs, I. Labadi and V. Izvekov, *J. Therm. Anal.*, 53 (1998) 343.
[116] B. Hariharanath, A.G. Rajendran, K.S. Chandrabhanu, M. Ravindran and C.B. Kartha, in: *Proceedings of the 3rd International High Energy Materials Conference and Exhibit HEMCE – 2000*, pp 469.
[117] Zhu Shun Guan, You Chen, Zhang Wen Yi and Mu Jing Yan,
[118] *Prop. Explos. Pyrot.*, 22 (1997) 36.
[119] D. Sheng and F. Ma, *Huogonpin*, 41 (1997) 36; CA 132 – 66235.
[120] S.A. Dehrle, *J. Energ. Mater.*, 15 (1997) 125.

INDEX

A

accumulation, 90, 106, 114, 134
accuracy, 9
acetic acid, 41, 50
acetone, 3, 120, 130
acid, viii, ix, 3, 4, 6, 7, 9, 17, 21, 22, 23, 40, 41, 42, 50, 51, 55, 56, 57, 58, 60, 61, 80, 81, 82, 85, 87, 88, 89, 91, 92, 114, 117, 120, 147, 150
acidity, 132
acrylonitrile, 2, 4, 87
activated carbon, 119, 120, 121, 123, 130
activation, ix, 117, 148
activation energy, 148
active site, 133
adaptation, 8
additives, 10, 53, 144
adenosine, 119
adenosine triphosphate, 119
adhesives, 14
adsorption, vii, ix, 1, 8, 9, 16, 26, 27, 28, 29, 34, 50, 102, 111, 117, 119, 120, 121, 123, 124, 125, 126, 127, 128, 130, 131, 132, 133, 134, 135, 136, 137, 139, 140
affect, 3, 19, 34, 50, 94, 104, 125, 127
agent, 31, 34, 66, 73, 80, 154
aging, 130
agriculture, 94
alcohol, 3, 77, 79, 88, 118
alcohols, 21, 118
algae, 119
alkylation, 144
alkylation reactions, 144
alternative, 56, 58, 71, 75, 87, 122, 150, 156
alternatives, 131
aluminium, 70, 130
aluminum, 24, 25, 29, 43, 50, 130
amendments, 90
amines, 60, 62, 66, 69, 129
ammonia, 19, 66, 147, 151, 154
ammonium, 56, 58, 62, 63, 65, 67, 80, 84, 126, 128
anaerobic sludge, 7, 17
ANC, 40
aniline, 154
animals, 90, 108, 109, 119
ANOVA, 92
APC, vii, viii, 39, 40, 41, 42, 43, 44, 46, 48, 49, 50, 52
aqueous solutions, ix, 69, 117, 123, 127, 132
aquifers, 18, 36
argon, 153
aromatic compounds, 16, 35, 126
aromatic hydrocarbons, 16, 120, 126
ascorbic acid, viii, 89, 92
ash, vii, viii, 39, 40, 41, 44, 45, 46, 48, 49, 50, 52, 53, 54, 120, 121
assessment, 8, 35, 54, 112
association, 101, 113
atoms, 80, 81, 118, 119, 122, 127
attention, 19, 71, 81
availability, 31, 90, 94, 106, 113, 114, 115
avoidance, 9

B

Bacillus subtilis, 13
bacteria, 11, 19
barriers, 3, 18, 128
BEA, 123
behavior, viii, ix, 39, 40, 46, 50, 52, 53, 54, 77, 117, 119, 122, 126, 127, 131, 132, 137, 140
benzene, 3, 8, 16, 88, 125
beverages, 4, 10
binding, 70, 124, 130
bioavailability, 90, 94, 112

bioconversion, 35
biodegradability, 11, 17, 21
biodegradation, vii, 1, 2, 7, 8, 16, 17, 18, 19, 26, 31, 32, 33, 34, 36, 128
biological processes, 19
biomass, vii, 1, 19, 21, 26, 31, 32, 33, 34, 122
bioremediation, 17, 18, 20
biosurfactant, vii, 1, 3, 21, 22, 33, 37
biosynthesis, 11
birds, 119
birth, 119
birth weight, 119
blood, 9
body, 7, 118
bonding, 143
bonds, 78, 129, 147
breakdown, 155
bromination, 56, 57, 60, 62, 66, 67, 82, 84
bromine, 58, 71
building blocks, 87
burning, 144
butadiene, 2, 4, 84
butadiene-styrene, 2, 4

C

cadmium, 37, 112, 113, 114, 115
calcium, 43, 50, 53
calibration, 148
Canada, 1, 4, 5, 6, 8, 18, 20, 21, 35
cane sugar, 121
caprolactam, 118
carbon, 3, 8, 19, 84, 118, 119, 120, 121, 128, 131
carbon dioxide, 19
carcinogen, vii, viii, 1, 2, 55, 80
cardiovascular system, 118
carrier, 56
catalyst, 3, 67, 132
catalytic properties, 144
cation, viii, 55, 56, 84, 113, 123, 124, 125, 126, 127, 128, 129, 130, 131, 132, 137, 140, 143
CEC, 106
cell, viii, 16, 55, 58, 82, 84
cell surface, 16
cellulose, 53
CH3COOH, 79
chemical properties, 21
chemical reactions, 133
chemical structures, 66
China, 42, 52, 55, 60, 70, 83
chlorination, 119
chlorine, 88, 118, 119, 121, 122, 127
chlorobenzene, 16
chloroform, 80, 118
chromium, 114
classes, 10
classification, 9, 10
cleaning, vii, viii, 10, 41, 43, 55, 81
cleavage, 7
CMC, 11, 12, 22, 23
CO2, 17, 19, 35, 41, 83, 84, 146, 147, 153, 155
coagulation, 128
coal, vii, ix, 7, 39, 40, 117, 118, 120, 121
cobalt, 150, 151, 153, 154
combustion, 22, 52, 53, 128, 129, 132
common symptoms, 109
community, 19
compatibility, 11
competition, 123, 124, 127
complex interactions, 133
complexity, 139, 154
components, 2, 4, 21, 22, 24, 34, 40, 42, 73, 111, 119, 144
composition, viii, 7, 11, 12, 21, 39, 40, 43, 49, 52, 73, 77, 79, 113, 114, 124, 130, 156
compounds, ix, 2, 3, 8, 11, 17, 18, 60, 94, 117, 118, 119, 120, 122, 124, 126, 127, 129, 135, 137, 139, 140, 144, 154, 155
concentration, vii, viii, ix, 1, 3, 7, 11, 12, 16, 17, 18, 20, 21, 22, 23, 24, 25, 26, 27, 29, 31, 32, 33, 34, 39, 40, 41, 42, 44, 47, 50, 56, 57, 58, 60, 61, 69, 80, 81, 82, 85, 89, 90, 92, 94, 104, 106, 107, 108, 109, 110, 111, 112, 113, 122, 123, 127, 130, 132, 133, 134, 135, 137, 138, 139, 144
concrete, 50
condensation, 75, 76, 78
conduct, 70
conductivity, 74, 144
configuration, 20, 131
conservation, 19
construction, 4, 8, 20
consumption, 108, 110
contact time, 24, 29, 31, 33, 34
contaminant, 4
contamination, ix, 3, 8, 16, 18, 34, 106, 117, 118, 128
control, vii, 39, 40, 41, 50, 52, 53, 54, 62, 94, 104, 107, 108, 109, 110, 127
conversion, ix, 69, 71, 87, 117, 118, 120, 124, 125, 150
cooling, 150
copolymers, viii, 2, 4, 55, 56, 67, 78, 79, 84
copper, 35, 113, 114, 115, 144, 145, 147, 148, 149
corn, 114
correlation, ix, 89, 90, 102, 103, 104, 106, 110, 111, 112, 135, 136, 137, 139

correlation coefficient, 102, 103, 104, 110, 135, 136, 137, 139
costs, 3, 18, 19, 20
coupling, 73, 75, 77, 86
coverage, 123, 127, 135
crops, 90, 106, 114
crude oil, 16, 37
curing, 77
cycles, 129
cycling, 112

D

danger, 83
death, 119
decomposition, ix, 19, 76, 108, 129, 143, 144, 145, 147, 148, 149, 150, 151, 153, 154, 155, 156
decomposition temperature, 76, 150
defects, 119
definition, 44
degradation, 6, 16, 17, 18, 19, 24, 32, 33, 35, 37
degradation process, 24
degradation rate, 33
dehydration, 145, 147, 148, 149, 153, 155
demand, 20, 24
Denmark, 114
density, 126, 127
depression, 110
derivatives, ix, 79, 87, 88, 117, 118, 119, 121
desorption, 2, 33, 128, 130, 134
destruction, 17
detection, 51
detergents, 7
deviation, 47, 137
dialysis, viii, 55, 56, 58, 80, 82, 85
dichloroethane, 71
diffraction, 147
diffusion, viii, 50, 55, 56, 85, 125, 133, 134, 139, 153, 154, 155
diffusion process, 50
diffusion rates, 125
diffusivity, 50, 54
digestion, 7, 19, 23
discharges, 118
discontinuity, 154
dispersion, 74, 86, 128
dissociation, 131, 151, 154, 155
distilled water, 20, 23, 24, 25, 26, 32, 92
distribution, 4, 40, 43, 94, 100, 101, 129, 131, 132
DMF, 69
dominance, 122
drinking water, 118
DRS, ix, 143, 150, 154
drugs, 62
dry matter, viii, 2, 8, 89, 106, 112
drying, 92, 125
DSC, ix, 143, 144, 148, 153, 154, 155
DTA curve, 148

E

East Asia, 5
Eastern Europe, 5
effluent, 2, 18, 34, 106, 120
elasticity, 75
electrical conductivity, 144, 145
electricity, 19
electrodes, 84
electroencephalogram, 9
electrolysis, 82, 85, 88, 130
electrolyte, viii, 55, 56, 58, 84
emergence, 92
emission, 40
endotherms, 150
energy, 2, 17, 19, 34, 129, 144, 154
environment, vii, 1, 2, 3, 6, 8, 16, 29, 32, 90, 94, 112, 120, 128
environmental impact, 119, 120
environmental protection, 128
Environmental Protection Act, 35
epoxy groups, 69, 70
equilibrium, 26, 27, 103, 125, 132, 133, 134, 135, 137, 139
equipment, ix, 83, 117
etching, 85
ethers, 71, 72, 86
ethyl acetate, 80
ethylene, 58, 74, 75, 77
Europe, 43
evidence, 154
evolution, 150, 155
exclusion, 127
experimental condition, 87
exposure, 9, 16, 24, 118
expression, 133
extraction, 90, 112, 113, 115, 120, 128

F

fabric, vii, viii, 39, 40, 41, 43, 54
fabrication, 4
fatty acids, 11
feedback, 40
fermentation, 21
fertilizers, 90, 94, 113

films, 80, 87
filtration, 42, 139
fish, 119
flame, 41, 144, 157
flocculation, 130
flotation, 10
flue gas, 40, 41
focusing, 16
food, 2, 4, 6, 9, 81, 106
Ford, 84
formaldehyde, 80, 88
formamide, 61
fossil, 52
France, 9, 84
free energy, 11
friction, 144, 156
fructose, 12, 147
fruits, 4
fuel, viii, 8, 19, 52, 55, 56, 58, 82, 84, 88, 120
functionalization, 84
funding, 34

G

gamma radiation, 84
gasification, 7
gasoline, 17, 18, 37
gastrointestinal tract, 118
gel, 73, 75, 77, 82
generation, 17, 19, 40, 87, 155
genetics, 11, 37
glass transition, 75
glass transition temperature, 75
glucose, 12, 147
glycol, 7
goals, 120
granules, 21
graphite, 3
grass, 8
gravity, 21
groundwater, vii, 1, 2, 3, 7, 9, 17, 18, 19, 30
groups, 9, 11, 43, 60, 61, 62, 64, 66, 67, 69, 74, 76, 80, 82, 87, 120, 124, 126, 144, 147
growth, 92, 103, 104, 105, 107, 109, 110, 119
growth rate, 119
guidance, 144

H

half-life, 7, 118
halogenation, 56
harm, 119
harmful effects, ix, 117
hazardous wastes, vii
hazards, 90, 112
health, vii, 112, 156
health problems, 156
heat, 144, 145, 149, 157
heat release, 144
heating, 19, 77, 129, 144, 145, 150, 153, 155, 156
heating rate, 77, 145, 150, 153, 155, 156
heavy metals, viii, ix, 2, 40, 41, 42, 89, 90, 91, 92, 93, 94, 100, 102, 103, 104, 105, 106, 107, 108, 109, 110, 111, 112, 113, 114, 115, 121
helium, 150
hexafluoropropylene, 80, 87
host, 129
House, 43, 91, 158
human exposure, 118
hybrid, 73, 74, 75, 76, 77, 78, 79, 82, 86
hydrazine, 156
hydrocarbons, 16
hydrogen, 23, 125, 126, 140
hydrogen bonds, 126, 140
hydrogen peroxide, 125
hydrolysis, 19, 75, 76, 78, 129
hydrophilicity, 60, 62, 124
hydrophobic properties, 126
hydrophobicity, 16, 123, 126, 140
hydroxyl, 7

I

identification, 151
immobilization, 90
impurities, 120, 125
incompatibility, 156
incomplete combustion, 17
incubation time, 114
India, 89, 91, 114, 143
indication, 27
induction, 154
induction period, 154
industrial wastes, ix, 117, 120
industry, 9, 17, 120, 121
influence, 78, 150, 151
inhibition, 21
innovation, 35
inoculum, 18
insulation, 2, 4
intensity, 27, 135
interaction, vii, 1, 16, 19, 70, 104, 126, 153
interaction effect, 104
interactions, 9, 47, 119, 120, 121, 126, 127, 128, 137
interference, 24

interval, 134, 145
ionization, 127
ions, 47, 50, 60, 130, 131, 143, 153, 157
IR spectroscopy, 156
iron, 3, 94, 101, 102, 103, 111, 112, 113, 147
irradiation, 62, 80
isomers, 127
isotactic polypropylene, 87
isotherms, ix, 27, 117, 135, 136
I-V curves, 85

J

Japan, 5
jobs, 74

K

kidneys, 118
kinetic parameters, 146, 149
kinetic studies, 134
kinetics, x, 130, 132, 133, 134, 143, 148, 155
knowledge, 137

L

labor, 139
lactose, 147
land, 2, 6, 8
landfills, 52
Latin America, 5
lead, viii, 39, 40, 42, 44, 45, 46, 47, 48, 49, 50, 51, 52, 54, 69, 72, 109, 114, 115, 156
lifespan, 119
ligands, 90, 144, 145, 148, 149, 152, 154
lipids, 13
liquids, vii, 3, 16, 119
location, 15
Louisiana, 8
low temperatures, 129
lubricants, 13
lysine, 13

M

magnesium, 122
management, 19
manganese, 102, 103, 111, 112, 113, 114
mannitol, 12
manufacturing, vii, 8
mass, 40, 42, 44, 82, 133, 134, 139, 145, 147, 148, 150, 152, 153, 155, 156
mass loss, 145, 147, 152, 153, 155
matrix, 50, 51, 73, 74, 77
measurement, 23, 54, 145, 152
measures, 23
mechanical properties, 129
mechanical stress, 155
media, 83
melting, 8
membranes, viii, 55, 56, 57, 58, 60, 62, 64, 65, 66, 67, 68, 69, 70, 71, 72, 73, 74, 75, 77, 78, 79, 81, 82, 83, 84, 85, 86, 87, 88
metabolism, 17, 37
metabolites, 17
metal extraction, 90
metal hydroxides, 130
metal salts, 90, 94
metals, viii, ix, 17, 50, 89, 90, 91, 92, 93, 94, 102, 103, 104, 106, 107, 108, 109, 110, 111, 112, 113, 115, 144
methanol, 3, 24, 25, 26, 32, 80
methyl groups, 82
Mexico, 4
microbial cells, 16
micrometer, 73
Middle East, 5
mining, 9, 35
Ministry of Education, 83
mixing, vii, viii, 39, 40, 41, 42, 46, 52, 73, 130, 133, 139
mobility, 90, 94, 102, 108
mode, 83, 137, 146, 150, 153
modeling, ix, 117, 132, 133, 137
models, ix, 7, 26, 117, 133, 136, 137, 139
modified polymers, 72
moisture, 17, 41, 91, 92, 156
moisture content, 91, 92
mole, 80, 147, 154
molecular structure, 125
molecular weight, 77, 120, 121, 126
molecules, 11, 119, 120, 121, 123, 124, 126, 127, 131, 140, 143, 144, 154, 155
monitoring, 8
monolayer, 124, 135
monomers, 69, 70, 71, 77, 79, 82
movement, 8, 29
mucous membrane, 9
mucous membranes, 9
multiple regression, 92
multiple regression analyses, 92

N

NaCl, 21, 22
nanocomposites, 75
NCS, 154, 155
needs, 22, 34, 127, 140
network, 73
neutral lipids, 11
nickel, 37, 85, 109, 114, 115, 154, 155
nitrobenzene, 126
nitrogen, 26, 80, 81, 82, 129, 132, 152
nitrogen gas, 26
NMR, 68, 79
nonionic surfactants, 130, 131
nucleation, 149
nuclei, 66, 67
nucleus, 149, 154
nutrients, 17, 90

O

observations, 101, 132, 145, 147, 150
occlusion, 102
OECD, 21
oil, ix, 2, 8, 12, 16, 19, 117, 118, 120
oils, 8, 11, 24, 118
optical microscopy, 151
optimization, 74, 87
organ, 127, 129, 136, 140
organic compounds, 2, 15, 16, 17, 118, 124, 126, 130
organic matter, 19, 22, 90, 101, 104, 106, 108, 111, 113
organic polymers, 87
Organization for Economic Cooperation and Development, 21
orientation, 124, 126, 127, 137
oscillation, 118
osmosis, 87
oxalate, viii, 89, 92, 148, 149, 154
oxidation, 4, 6, 120, 144, 145, 154
oxides, 94, 101, 102, 103, 144, 156
oxygen, 3, 17, 18, 19, 20, 24, 26, 29, 31, 34, 122, 144, 153
ozone, 7

P

PAA, 6, 7
packaging, 2, 4
paints, 9, 14
parameter, 32, 120, 123, 127, 133, 135, 136, 137, 140, 149
particles, 73, 94
partition, 5, 8, 90, 119, 126
pathways, ix, 6, 143, 145
peat, 18, 36, 120, 121
permeability, 63
permeation, 62
permit, 62
peroxide, 3, 4
perspective, 83
pH, viii, 11, 12, 19, 21, 40, 41, 42, 44, 45, 47, 50, 51, 77, 89, 91, 92, 93, 94, 102, 103, 106, 108, 110, 111, 112, 114, 115, 121, 122, 123, 124, 127, 130, 131, 136
pharmaceuticals, 9, 118
phase transformation, 3, 18, 20
phenol, ix, 117, 118, 119, 120, 121, 123, 125, 126, 127, 128, 129, 131, 134, 136, 137, 139
phenolic resins, 118
phosphates, 130
phospholipids, 11
phosphorylation, 119
physical properties, 12, 118
physical treatments, 2, 17
physicochemical properties, 118, 140
physiology, 37
plants, viii, 7, 19, 39, 40, 41, 47, 52, 90, 92, 103, 106, 107, 108, 109, 110, 111, 112, 113, 114, 115, 118, 119, 121, 133
plasma, 41, 80, 124
plastics, 2, 4
platinum, 84
PM, 13
polarity, 127, 129
pollutants, ix, 106, 117, 120, 121, 128, 130, 139, 144
pollution, vii, 39, 40, 41, 50, 52, 53, 54, 120
poly(vinyl chloride), 87
polycarbonate, 92
polychlorinated biphenyls (PCBs), 17
polymer chains, 77
polymer films, 80, 87
polymeric chains, 73
polymerization, 3, 10, 70, 74, 86, 87, 124, 150
polymers, 6, 7, 35, 56, 62, 66, 68, 71, 72, 82, 84, 86, 87, 114
polypropylene, 80, 81
polystyrene, 2, 4, 6, 18, 24, 56, 66, 67, 80, 84, 85, 88
polythene, 91
polyurethane, 120
polyurethane foam, 120
polyvinyl chloride, 80
pools, 94, 100, 101, 112, 114
poor, 27, 94, 133, 135, 136
population, 3

porosity, 87, 120
positive correlation, ix, 89, 102, 111, 112
positive relation, 103
positive relationship, 103
potassium, 23, 50, 156
power, 8, 121
power plants, 121
precipitation, 94, 101, 102, 111
prediction, viii, 39, 40
preference, 137
preparation, viii, 55, 56, 67, 70, 71, 73, 74, 75, 80, 82, 83, 84, 85, 86, 87, 121, 127, 128, 129, 130, 131, 132, 140, 144, 150, 157
pressure, ix, 62, 117
prevention, 128
probability, 92
probe, 41, 42, 131
production, 2, 4, 5, 11, 19, 81, 82, 112, 118, 154
program, 83
propane, 70, 71
propylene, 4
Pseudomonas aeruginosa, 12, 13
PTFE, 80
pulse, 118
pure water, 81
purification, 9, 41, 42, 125
PVA, 74
PVC, 80
pyrolysis, 120

Q

quaternary ammonium, viii, 55, 56, 64, 66, 67, 124, 126

R

radiation, 87
radioactive waste, vii
rain, 51
range, 9, 74, 75, 76, 77, 107, 109, 118, 122, 124, 132, 145, 150
rape, 114
raw materials, 9, 121, 157
reaction rate, 151
reaction temperature, 67, 68, 130, 150, 151
reagents, 71, 81, 86, 88
recovery, 2, 9, 10, 19, 58, 60, 80, 82, 83, 85, 87, 128
recycling, 17
redistribution, 101, 102
reduction, 9, 74, 109, 127, 153
regeneration, 119, 120, 121, 123, 128, 129, 132
Registry, 37
regression, 45
regulation, 44
rehydration, 129
relationship, 7, 26, 45, 102, 113, 133
relaxation, 155
replacement, ix, 117, 129, 157
reserves, 72
residues, viii, 35, 39, 40, 41, 42, 43, 44, 45, 46, 49, 50, 52, 53, 54, 144, 149
resins, 2, 4, 66, 67
resistance, 58, 60, 69, 72, 119, 127, 133, 139
respiration, 9, 21, 118
respiratory, 9, 84
retention, 29
rhamnolipid, vii, 1, 3, 12, 16, 19, 21, 23, 25, 31, 32, 33, 34, 37
rice, 120
rice husk, 120
risk, 3, 18, 20
room temperature, 41, 150
rubber, 24, 25, 29, 70

S

safety, 35
salinity, 11
salts, viii, ix, 50, 80, 84, 89, 90, 91, 93, 94, 95, 96, 97, 98, 99, 101, 102, 103, 104, 105, 106, 107, 108, 109, 112, 113, 120, 144, 149, 150
sample, 23, 24, 25, 41, 42, 148, 150
sampling, 52
saturation, 91, 130
sediment, 17
sedimentation, 128
sediments, 2, 8, 17, 101, 113
selecting, 72
selectivity, 11, 60, 74, 80
self, 32, 33, 144
senescence, 109
sensing, 56
sensitivity, 50, 156
sensors, 56, 144
separation, 57, 74, 81, 87, 125, 127, 129, 138
series, viii, 18, 19, 39, 40, 41, 43, 46, 50, 52, 56, 57, 58, 60, 62, 65, 77, 80, 81, 85, 150
sewage, 91, 101, 106, 112, 113, 114, 115
shape, 127
shock, 144
SII, 24
silane, 73, 77, 80, 86
silica, 73, 75, 77, 121
silicon, 50

silver, 23
SiO2, 40, 42, 43, 44, 45, 46, 49, 52, 74, 75, 86, 122
sites, 2, 8, 82, 127, 129, 130, 131, 132, 137, 155
skin, 118
sludge, viii, ix, 2, 18, 19, 21, 89, 90, 91, 92, 93, 94, 95, 96, 97, 98, 99, 100, 101, 102, 103, 104, 105, 106, 107, 108, 109, 110, 112, 113, 114, 115, 121
sodium, 50, 62, 63, 120
soil, vii, viii, ix, 1, 2, 3, 8, 13, 16, 17, 18, 19, 20, 22, 24, 25, 26, 27, 28, 29, 31, 33, 34, 35, 36, 37, 50, 54, 89, 90, 91, 92, 94, 101, 102, 103, 104, 106, 107, 108, 109, 110, 111, 112, 113, 114, 115, 118, 128
sol-gel, 73, 74, 75, 76, 77, 78, 79, 82, 86
solid phase, 114
solid state, 145
solid surfaces, 130, 131
solid waste, vii, 8, 17, 39, 40, 41, 52, 53, 54
solubility, 9, 16, 50, 60, 114, 118, 121, 126
solvents, 9, 60, 69, 128
sorption, 16, 26, 27, 34, 35, 40, 49, 50, 121, 126, 127, 133, 134, 136, 137
sorption process, 40, 134
South Asia, 5
Soxhlet extractor, 128
species, 19, 41, 50, 81, 106, 111, 124, 128, 144, 145, 156
specific surface, 130, 144
spectrophotometry, 92
spectroscopy, ix, 143, 150
spectrum, 144, 154
speed, 41
spleen, 118
stability, 58, 59, 78, 80, 85, 114, 151, 156
stabilization, 52
stages, vii, 1, 58, 132, 138, 139, 145, 153
steel, 18, 121
steel industry, 121
storage, 139
streams, vii, 39, 40, 139
strength, 69, 73
strong interaction, 126
structural changes, 129
students, 83
styrene, vii, viii, 1, 2, 3, 4, 5, 6, 7, 8, 9, 16, 17, 18, 19, 20, 24, 25, 26, 27, 28, 29, 31, 32, 33, 34, 35, 36, 37, 55, 56, 67
styrene-butadiene rubber, 2, 4
substitution, 56, 57, 59, 67, 68, 69, 72, 88, 118, 150
substrates, 11, 12, 16, 72, 126
sulfur, 53, 147
sulfuric acid, viii, 23, 55, 56, 85, 122
Sun, 157
superiority, 129
supply, 3, 20, 51, 106
surface area, 119, 120, 121, 127, 128, 129, 131, 132
surface properties, 119, 126, 137
surface structure, 127, 129, 131, 140
surface tension, 11, 12, 22, 23
surface treatment, 123
surfactant, vii, 1, 10, 11, 12, 13, 16, 19, 123, 128, 130, 131, 132, 140
suspensions, 91
sustainable development, 19
swelling, 70, 74, 122, 128, 129, 131
symmetry, 149
synthesis, 3, 4, 7, 35, 72, 83, 88, 119, 125
systems, 3, 16, 17, 19, 43, 53, 56, 121, 127, 134, 139

T

Taiwan, 39, 40, 43, 44, 52
TDI, 74
technology, 2, 3, 17, 34, 84
temperature, ix, 12, 16, 18, 32, 34, 41, 58, 59, 60, 63, 67, 69, 73, 76, 77, 79, 82, 85, 88, 113, 117, 118, 125, 127, 129, 143, 144, 145, 146, 150
tension, 2, 11, 21, 23
tetrahydrofuran, 61
textiles, 13
TGA, 76, 77, 79
thermal analysis, 149
thermal decomposition, ix, 143, 147, 148, 149, 153, 154, 155, 156
thermal degradation, 79
thermal properties, 75, 145, 147, 150
thermal stability, ix, 75, 77, 129, 132, 143, 144, 149, 155
thermal treatment, 132
thermograms, 77, 145, 148, 149
thermolysis, ix, 143, 144, 145, 147, 149, 153, 154, 156
time, viii, 17, 18, 26, 30, 31, 34, 39, 40, 52, 58, 59, 60, 69, 77, 94, 118, 125, 127, 132, 133, 134, 139, 147, 154
tissue, ix, 89, 90, 92, 107, 108, 109, 110, 111, 112, 113
titanium, 78, 85
toluene, 8, 36, 126
toxic effect, 119, 120
toxic products, 17
toxicity, vii, 11, 21, 35, 39, 40, 106, 108, 109, 119
TPA, 61, 92
trace elements, 112
transactions, 85
transformation, 7, 90, 101

transformation processes, 7
transformations, 155
transition, ix, 77, 143, 144, 147, 150, 155
transition metal, ix, 143, 144, 147
transport, 7, 16, 56, 83, 84, 133
transportation, 20
trend, 60
turbulence, 7
Turkey, 117

U

ultrasound, 124
uncertainty, 135
uniform, 119
United States, 2, 4
universe, vii
upper respiratory tract, 9
UV, 24, 25, 26, 80

V

vacuum, 150, 151
valence, 58
values, 47, 60, 62, 69, 76, 103, 106, 123, 127, 130, 131, 136, 137, 148, 149, 150
variable, 130
variables, 134
variance, 92
variation, 18, 79
vegetables, 4, 107, 114
versatility, 73
vertigo, 118
vinyl chloride, 87
vinyl monomers, 79, 82
volatility, 3
volatilization, 7, 26, 147

W

washing procedures, 2
waste disposal, 19
waste incineration, 40
waste treatment, 2, 17, 34, 90, 118, 121
waste water, 83
wastewater, 7, 18, 19, 21, 119, 122, 127, 131
water, viii, 2, 3, 7, 8, 14, 17, 18, 21, 23, 33, 42, 56, 57, 58, 60, 61, 62, 63, 66, 69, 70, 74, 77, 81, 82, 86, 89, 91, 92, 94, 101, 102, 103, 104, 106, 109, 112, 118, 119, 121, 123, 124, 126, 127, 128, 130, 140, 147, 150
weakness, 118
weight loss, 77, 78, 79
Western Europe, 5
wood, 118
words, 119, 135, 139
work, 8, 9, 11, 50, 91, 118, 127, 129, 130, 134, 135
workers, 9, 94, 102, 145, 147, 155

X

XRD, 124, 132, 149

Y

yeast, 11, 22
yield, viii, ix, 19, 89, 92, 106, 112, 114, 115, 131, 151, 155

Z

zeolites, 123, 132
zinc, viii, 39, 110, 113, 114, 115, 156
zinc sulfate, 114
ZnO, 144, 155, 156